APHRODISIACS — THE SCIENCE AND THE MYTH

APHRODISIACS
THE SCIENCE AND THE MYTH

P.V. TABERNER

CROOM HELM
London & Sydney

© 1985 Peter V. Taberner
Croom Helm Ltd, Provident House, Burrell Row,
Beckenham, Kent BR3 1AT
Croom Helm Australia Pty Ltd, First Floor, 139 King Street,
Sydney, NSW 2001, Australia

British Library Cataloguing in Publication Data

Taberner, P.V.
    Aphrodisiacs.
    1. Aphrodisiacs
    I. Title
    615'.766   HQ12

    ISBN 0-7099-2477-1

Typeset by Mayhew Typesetting, Bristol, UK
Printed and bound in Great Britain
by Billing & Sons Limited, Worcester.

CONTENTS

FIGURES

PREFACE

The planning and writing of this book has taken rather longer than I had originally intended; what began as a modest literary project for two second-year medical students has expanded over eight years to become a complete book. The subject matter lent itself all too easily to a sensationalist approach yet, on the other hand, a strictly scientific approach would probably have resulted in a dull dry text of little interest to the general reader. I have therefore attempted to bridge the gap and make the book intelligible and entertaining to the non-specialist, but at the same time ensuring that it is factually correct and adequately researched for the scientist or clinician.

I have always been impressed by Sir J.G. Frazer's introduction to his classic book *The Golden Bough* in which he apologizes for the fact that an article originally intended merely to explain the rules of succession to the priesthood of Diana at Aricia had expanded, over a period of thirty years, to twelve volumes. The present work cannot pretend to such heady levels of academic excellence.

In spite of this expansion there are areas upon which I have only been able to make the briefest comment. There was no room for Hoffman's Water of Magnanimity or the winged ant sweetmeats of the harems of Arabia. An entire book could have been devoted to the subject of alcohol, the universal social lubricant, and certainly the most widely used aphrodisiac today. New drugs of abuse have probably become fashionable since Chapter 7 was written. New case reports of drugs with unexpected aphrodisiac or anaphrodisiac side-effects will most likely have appeared in the medical literature. New pseudo-aphrodisiacs have certainly appeared in the ever-increasing number of hard-core pornography magazines. In fact, I do not imagine for one moment that the present work represents the last word on the subject; the legacy of Aphrodite is still too active an influence on our social and marital habits for that.

A large number of individuals have contributed directly or indirectly to the writing of this book, but I am particularly indebted to Boz and Judie, who started it all; Jeannie, Pat and Erica for typing early drafts of the text; to Jeanette Edwards for her excellent artwork; Phil Darby for his fine photographic work and assistance with the figures; Vivian Brown for proof-reading the quotations in French; Professor Burrow for translating the Anglo-Saxon text; Dr Roger Price of the Wellcome Museum of the History of Medicine and to Dr Philip Brown and Professor Gerald Kerkut, who provided essential advice and enthusiasm at key moments.

Finally, I thank my wife, who provided invaluable practical support and help throughout the protracted gestation of the book and to whom it is dedicated, with much love.

1 INTRODUCTION: THE NATURE OF APHRODISIACS

Aphrodite, the Greek goddess of desire, supposedly sprang from the seed of the god Uranus, symbolized by the foam on the sea from which she rose, naked, near the shores of Cythera. She has since been worshipped as a fertility goddess in shrines throughout the countries of the mediterranean area. To the Romans she was Venus, the irresistibly beautiful goddess of love. During a promiscuous existence on Earth she had, according to Greek legend, sons by various partners and named, appropriately enough, Eros, Priapus and Hermaphrodite. Even if the goddess and her family have long since passed into mythology, she has left a more tangible legacy in the form of a host of animal and vegetable products bearing her name and reputedly capable of facilitating sexual desire.

In spite of the fact that aphrodisiacs have featured in art and literature throughout history, modern medical and scientific opinion has tended to dismiss them and conclude that no such substance, in the classical sense, can exist. It is this strict definition of an aphrodisiac as a drug, or other substance, capable of inducing venereal desire or lust, which is perhaps responsible for this rather disappointing conclusion. However, it has been suggested by Gawin[1] that, in view of the contemporary widespread recreational use and abuse of drugs in a sexual context, the definition should be enlarged to include effects upon the subjective pleasure experienced during sexual activity. This seems a most reasonable idea in the light of the current popularity of both legal and illegal drugs among large sections of the community. It would also suggest one very good reason behind the popularity of these recreational drugs. This broader definition of aphrodisiac has therefore been adopted in the selection of material in this book.

It is very easy merely to claim that a substance has aphrodisiac properties; the practical proof of the claim at a level that would satisfy a sceptical member of the scientific or medical profession is far more difficult. The quantitative measures of sexual activity such as frequency of coition and ejaculation or orgasm are more appropriate to the animal laboratory than the bedroom, although they are functions that lend themselves to accurate measurement. However, sheer physical performance and stamina are probably less important to the individuals who do

1

not consider themselves sexually inadequate and use aphrodisiacs solely as a means to increase their pleasure. On the other hand, there are some people who do suffer from various sexual difficulties which may have a physical or mental basis, and there are clinically effective drugs available for that treatment. On a lighter level there are a vast range of cosmetic preparations which are used consciously or unconsciously as aphrodisiacs. In the UK there is a range of perfumes and toiletries sold under the brand name 'Aphrodisia', but far more subtle means are used to market such products.

It is therefore too easy to dismiss aphrodisiacs as a myth or a scientific impossibility. The historic search for the love drug, which has occupied every civilization throughout history, is not merely the fruitless quest for the elixir of life or the philosopher's stone which would turn base metals into gold. In contrast, it may have some basis in fact. Incidentally, the elixir of life, which was so industriously sought by the medieval alchemists, was also reputed to possess aphrodisiac properties.

This work is an attempt to cover not just those substances which influence sexual performance, but also the substances which have been used to procure love and sexual attraction. There is therefore a collection of love charms and potions described, which, although virtually extinct in Western Europe, are still an important feature of life in less industrialized communities. Since one of the strongest motives for the use of aphrodisiacs is a feeling of sexual inadequacy, some attention has been given to describing the specific problems for which current medical practice can provide effective remedies. Although the drugs used clinically for the treatment of sexual disorders are not necessarily aphrodisiac in themselves (they have little or no effect in normal healthy individuals), they can nevertheless produce striking improvements in cases where there is some underlying physical basis for the sexual problem.

Finally, having examined the history and basis of aphrodisiacs, it seems only fair to consider the substances which have the opposite effect, the anaphrodisiacs. This is a particularly relevant subject at present since several medically prescribed drugs have marked anaphrodisiac side-effects. For many people, a sudden unaccountable loss of libido is something about which they can feel too embarrassed to consult their doctor, whereas it can often be a temporary problem associated with the drugs which the same doctor is prescribing for an unrelated illness. There are also some drugs which can be used for their specific anaphrodisiac effect to diminish an excessive libido where this seems appropriate.

The books written in the past on the subject of aphrodisiacs have invariably been totally uncritical of the material they describe, in some cases merely providing a catalogue in which the unlikely and the absurd appear alongside potentially active products. It is therefore my hope that the present book will help to rationalize our ideas concerning the use of aphrodisiacs, and also discourage the promiscuous use of inherently toxic substances or expensive quack remedies of no intrinsic worth.

What are Aphrodisiacs?

The *Oxford English Dictionary* defines an aphrodisiac as 'A drug, etc. inducing venereal desire'. Venereal desire can probably be more readily appreciated as a desire for sexual stimulation, this desire often being described as the sexual appetite. However, an increase in appetite is not necessarily associated with a corresponding increase in the ability to satisfy that appetite. Anyone who has used alcohol to release their inhibitions and increase their sexual confidence only to find that either they or their partner has reached a level of intoxicated incapability can bear witness to this point.

We can therefore regard alcohol as an example of the type of aphrodisiac that can act specifically to increase the sexual appetite, but a second category of aphrodisiac can exist which acts to increase the ability to indulge in sexual activity. Drugs of this latter type may have a use in clinical medicine to restore the capability of a patient who has suffered from some pathological disorder which has resulted in organic impotence. The best example of this type of drug is the use of testosterone replacement therapy in cases where accidental or surgical castration has occurred.

The third potential category of aphrodisiacs consists of those substances which act to increase the capability or prolong the ability to maintain successful sexual activity. The best known example is probably illustrated by the widespread use of local anaesthetic creams on the penis in order to prevent premature ejaculation. *Ejaculatio praecox* is quite a common disorder often self-treated by use of some fairly expensive proprietary remedies which are marketed in effect as aphrodisiacs.

Thus, although in theory an aphrodisiac is strictly an exciter of lust, in practice anything which, by any means, increases the capacity for sexual enjoyment will tend to increase the appetite and can be termed an aphrodisiac. Drugs certainly exist which can restore a diminished sexual ability to normal in so far as 'normal' can be defined for any particular individual. Also, there have been many reports of varying degrees of scientific reliability in which a medically prescribed drug has been found to cause an increase in the libido of the patient. In many cases it has not been the patient who has complained of this side-effect to the doctor who prescribed the drug but the patient's partner on whom the increased libido has been exercised.

An interesting situation occurs rarely in one form of temporal lobe epilepsy. Epileptics can very often predict the onset of a fit by the occurrence of a hallucinatory or psychic state or even by just an odd taste in the mouth. In rare cases the patient has reported having sexual fantasies of considerable erotic potency. Not unexpectedly these patients have often been reluctant to accept any form of treatment for their epilepsy.

When we come to the question of whether drugs exist which can specifically increase the desire, we are on far more difficult ground. How can the level of desire be accurately measured? How certain can we be that the level of desire has not been fortuitously affected by some outside factor such as the menstrual cycle or the diet? Recently, several physiological measures have been developed to assess objectively the level of sexual arousal (see page 164), but sexual desire can really only be assessed subjectively by the individual and here conditioning will be an important factor.

Sexual Conditioning

Classical conditioning was first demonstrated by the Russian physiologist Pavlov. In his experiments dogs were fed at specific times. The appearance of the food and its aroma produced the perfectly natural, or unconditioned, response of salivation. However, in these experiments a bell was rung immediately before the food was presented. The dogs therefore came to associate the ringing of the bell with the certain knowledge that food was in the offing. Eventually, the ringing of the bell alone, even outside normal feeding times, was sufficient to provoke salivation in the absence of any food. The bell was the conditioning stimulus and the salivation now became the conditioned response.

If we now apply this technique to the vastly more complicated human situation, we can still observe classical conditioning. If a successful sexual encounter is regularly preceded by a candlelit dinner in intimate surroundings, then the conditioning stimuli of food, soft lights and music may eventually be sufficient to produce sexual arousal. Conversely, the association of an unpleasant or embarrassing event with an early sexual encounter can produce severe problems in later sexual activity. This aversive conditioning has actually been employed in attempts to treat patients who have shown criminal sexual tendencies and in some cases has been successful. Human sexual arousal is a complicated phenomenon, but simple conditioning situations of the type described can probably be just as potent as aphrodisiacs or anaphrodisiacs as any drug.

The Inequality of the Sexes

The basic physiological differences between the two sexes can result in an aphrodisiac being effective in the one but anaphrodisiac in the other. The human animal is unusual in that the female is able to accept the male at any time during the menstrual cycle although ovulation only occurs once a month, about fourteen days before menstruation. In most species the female only shows an interest in the male at the time of ovulation when she is on heat and likely to conceive.

The question of whether or not women experience variation in their level of desire during the menstrual cycle has provoked much research. A well-controlled study,[2] in which 60 North American married white women contributed information anonymously each day on whether they had coitus, orgasm or menstruation, yielded some interesting results. The rates of coitus and orgasm coincided closely and showed two very prominent peaks at the middle of the cycle 14 to 18 days before the period of menstruation) and again at the end of the cycle 3 to 1 days before the period). The first peak clearly coincides with the time of ovulation and suggests that some intrinsic hormonal change may well affect the libido at this time. The second peak is more difficult to explain although the cause may be the result of the hormonal changes that are also responsible for pre-menstrual tension. The increase in sexual desire reported to occur immediately following menstruation is most likely to be due to enforced abstinence during the time of menstruation. (The endocrinological changes that occur during the menstrual cycle are shown in Figure 6.3.)

From a purely biological point of view the only requirements for a successful sexual encounter are that the male should be able to achieve penetration and ejaculation. From a social point of view both partners are looking for sexual pleasure and satisfaction so the male orgasm and ejaculation should coincide with an orgasm or climax in the female. To achieve this reliably requires careful timing and a good deal of experience with the same partner. It is to these ends, namely the delaying of the male orgasm (after which the erection is often lost) and the encouragement of the female orgasm, that the use of aphrodisiacs is mainly directed. The plan is therefore to increase the sensitivity of the clitoris in the female and at the same time decrease the sensitivity of the penis. From a biological point of view we are designed so that the male climaxes before the female in order to propagate the species. If the female was the first to be satisfied, then the chance of getting pregnant would be considerably reduced. Women are biologically disadvantaged in that they are so often left consummated but unsatisfied. On the other hand they have the considerable advantage of being able to achieve multiple orgasms and probably more pleasure than the man. It is only civilized man who has the time and opportunity to treat sex as a pleasurable activity rather than as an essential act to ensure the propagation of the species. It is not just this quest for pleasure, however, that has led man in all parts of the world to seek for drugs, potions, elixirs and charms to increase his sexual activity. Sexual potency is part of the male ego, and the anxiety and loss of face that are associated with a declining sexual prowess are common to all cultures. This is why aphrodisiacs feature throughout recorded history and show little sign of decline in popularity in the present day.

The object of this book is not merely to list the myriad substances that have been used as aphrodisiacs, but to explain the rationale behind their usage and assess their true effectiveness. There are still several frankly dangerous substances used today with reputations as aphrodisiacs that are based largely on mythology. It is the separation of the truth from the myth that I have attempted in this book.

Aphrodisiacs and the Law

Many of the reputed aphrodisiacs described in this book are drugs or poisons. In fact, there is no real distinction between a drug and a poison; many clinically used drugs can produce toxic symptoms at doses only slightly above the recommended therapeutic dose. There is also a widespread and unfortunate misconception that if a little bit of something does you good, then a little bit more should be even better. It is quite likely that many cases of accidental self-poisoning arise from this erroneous idea. The law is concerned with drugs and poisons within two main areas: first, the food, drugs and cosmetics legislation, which seeks to ensure that proprietary products are safe for public use, and secondly, the misuse of drugs legislation, which makes the selling or even the possession of certain specific drugs a criminal offence. The latter is aimed at preventing the unauthorized use of drugs for which there is reason to believe that they possess a potential for abuse or addiction. Alcohol and tobacco, although they clearly satisfy both criteria, also yield enormous sums of money to the exchequer from the high rates of tax levied on them, and they are never likely to be banned by law.

United Kingdom Law

There is no statutory definition of a poison, and so the law can be fairly flexible on the subject. The first example of drugs legislation in the UK was the Arsenic Act of 1851, but the Offences against the Person Act of 1861 referred specifically to a 'poison or other noxious thing'. A noxious thing can obviously be anything that produces unpleasant or toxic effects. The scientific definition of a poison, given in the *Encyclopaedia Britannica*, is: 'a substance which by its direct action on the mucous membrane, tissue or skin, or after its absorption into the circulatory system can, in the way in which it is administered injuriously affect health or destroy life'.

A drug is far less easy to define, since it should include substances used as placebos (see page 167), which have no intrinsic therapeutic efficacy, but which nevertheless can act in a beneficial way, perhaps by suggestion. Placebos normally consist of glucose, lactose or chalk. Vitamins and other dietary supplements can also be prescribed as drugs, although we would not normally think of them as such. From the judicial point of view, Lord Goddard, when he was Lord Chief Justice in 1957, defined a drug as follows: 'A drug means a medicine or medicament, something given to cure, alleviate or assist an ailing body.' This is the definition likely to be accepted in an English court of law.

Most of the substances or mixtures described in the following chapters will be subject to the laws regarding poisoning. The administration of poison is a criminal offence whether it is with the express intent to commit murder or merely to cause injury. It is also an offence to administer a poison or stupefying drug in order to facilitate the committing of a crime. This is singularly relevant in the context of aphrodisiacs since there have been many cases, both in the UK and abroad, in which the intended crime has been unlawful sexual intercourse. Section 23 of the Offences against the Person Act of 1861 states:

'Whosoever shall unlawfully and maliciously administer, or cause to be administered to or taken by any other person any poison or other destructive or noxious thing, so as thereby to endanger the life of such person, or so as thereby to inflict upon such person any grievous bodily harm, shall be guilty of felony.'

Grievous bodily harm, by the way, is the rather archaic English way of defining any act which seriously interferes with health or comfort, it is not restricted merely to physical violence. These legal phrases may seem stilted, but they effectively cover almost any eventuality. It is not even necessary for a 'noxious thing' to be dangerous to life, or for its effects to be permanent: if a case comes to court, it is the members of the jury who have the task (with helpful directions from the judge) of ultimately determining whether the circumstances of that case actually constitute grievous bodily harm. The indiscriminate use of Spanish flies today would almost certainly come into the category of causing grievous bodily harm; they are very noxious things indeed.

It is a lesser offence to administer poison with intent to injure, aggrieve or annoy. In the last century, when Spanish flies (cantharidin) were very much more readily available than they are today, several cases of malicious poisoning came to court. If the victim died, the charge was manslaughter, but in one case (*Regina* v. *Hennah*, 1877) the accused administered only a small dose which caused no apparent harm, and an acquittal was obtained by virtue of the fact that it could not be proved that there was any intent to harm or cause injury on the part of the defendant. Although there is more sexual freedom today compared with a hundred years ago, there is certainly less freedom to administer potentially toxic substances, for whatever purpose.

Although the use of sedative or stupefying drugs is covered by Section 22 of the Offences against the Person Act, the use of any drug to facilitate unlawful sexual intercourse is specifically covered by Section 4 (1) of the Sexual Offences Act of 1956:

'It is an offence for a person to apply or administer to, or cause to be taken by a woman, any drug, matter, or thing with intent to stupefy or overpower her so as thereby to enable any man to have unlawful sexual intercourse with her.'

Anyone using a purported aphrodisiac in order to further his ends with an otherwise unwilling partner obviously leaves himself open to prosecution under one or more of these Parliamentary Acts. It is hoped that the seeker after the true aphrodisiac, after reading the following chapters, will be less inclined to experiment with noxious substances of very doubtful efficacy.

The Misuse of Drugs

Very many of the drugs which are currently abused by sections of the population are employed for, among other things, their supposed aphrodisiac effects. These drugs will be considered in detail in Chapter 7, but for the moment it is worth examining the laws that relate to the recreational abuse of such substances.

Drug abuse and addiction are not new, but relevant legislation to control the distribution of dangerous drugs was only introduced in England in 1917 in response to the epidemic of cocaine abuse resulting partly from its widespread use among the armed forces in the First World War. (A similar problem arose with amphetamine in Japan immediately following the Second World War.) In 1920 the Dangerous Drugs Act prohibited the import and export of opium (which, incidentally, had been freely available over the chemist's counter until 1868.) The import, export and production of, and dealing in, cocaine, heroin, morphine and raw opium were also subjected to strict controls. As newer drugs, such as the barbiturates, amphetamines and cannabis, became available, fresh legislation was enacted in an attempt to limit their misuse. However, the various relevant Acts of Parliament, from 1923 to 1967, were proving completely inadequate and were eventually replaced by the Misuse of Drugs Act of 1971. This made new provisions with respect to dangerous or otherwise harmful substances and created a list of those substances that were to be controlled. New drugs can be added to this list without recourse to fresh legislation and the delays that it entails. The very large majority of the drugs described in Chapter 7 are 'controlled' drugs within the meaning of this Act and their sale or possession could result in prosecution.

Both Britain and the United States have international obligations as parties to the Single Convention on Narcotic Drugs of 1961, but no laws have yet been designed to cope successfully with the flood of newly synthesized novel compounds with psychotomimetic effects and consequent potential for abuse. The recent relaxation of the law in Spain governing the possession of cannabis for personal use is just one indication of how difficult international control of drug abuse has become. Customs seizures of narcotics (heroin, morphine and opium), cocaine, marijuana and cannabis resin continue to increase, implying an underlying growth in the scale of use of these drugs.

United States Law

Food and drugs legislation is somewhat different in the US, being complicated by the variations in law between the various states. The principal relevant federal laws (which apply to all states) are the Food, Drug, and Cosmetic Act of 1938, and the Uniform Narcotic Drug Act 1973. The 1938 Act updated the original Food and Drug Act of 1906. The original legislation was introduced as a result of the campaigning efforts of H. Wiley, and in response to the increasing number of complaints from dissatisfied customers of the legions of quack doctors who flourished at that time. The principal aim of the Act was to protect customers from 'misbranding' and the fraudulent claims of manufacturers with regard to the efficacy of their secret remedies. Many of these preparations were for sexual weakness and some specific cases are described in Chapter 4. Prior to the 1906 Act, only mail order companies could be taken to court. The sections of the United States Revised Statutes dealing with the postal laws made it an offence to 'conduct a scheme or device for obtaining money or property through the mails by means of false or fraudulent pretenses, representations or promises'. In practice, this statute was used successfully to bring to court the purveyors of remedies claimed to have aphrodisiac or sexually invigorating properties. More recently the same laws were employed to prevent the mail marketing of Pega Palo (see page 92).

In the not too distant past, the US attempted to eliminate drug abuse by far more rigorous legislation than has ever been considered in the UK. The 18th amendment to the US Constitution (the highest body of US legislation), which was adopted on 29 January 1919, sought to prohibit the manufacture, sale or transportation of intoxicating liquors for beverage purposes within all territory subject to the jurisdiction of the Federal Government. The experiment was a complete failure, the law was widely flouted, and the era of prohibition was ended by the passing of the 21st amendment in December 1933 which repealed the original amendment. Once any drug has become established in a society, it seems to be almost impossible to eliminate it by legislation.

Under the terms of the Uniform Narcotic Drug Act it became unlawful for any person to manufacture, possess, have under his control, sell, prescribe, administer, dispense, or compound any narcotic drug, except as authorized in the Act. Narcotic drugs, as originally defined, were taken to include coca leaves (the source of cocaine), opium, cannabis, and every substance neither chemically nor physically distinguishable from them, and any other drugs whose importation or possession was prohibited, regulated or limited under the federal narcotic laws existing on the date of the event to which the act was to be applied. Like the Misuse of Drugs Act in the UK, this represented an attempt to keep up with new developments on the drugs scene without the necessity for constant additions to the existing legislation.

Advertising Aphrodisiacs

Before literacy became widespread and journals and newspapers existed, advertising medicines and potions was limited to word-of-mouth recommendation or 'barking', much in the way that a travelling fair or market would attract customers today. In situations where a significant proportion of potential customers might be literate, pamphlets or broadsheets could be distributed or posted on walls to advertise a forthcoming lecture by 'Doctor' Dogood or 'Professor' Nostrum. The practical side of selling remedies for male weakness or female complaints would rely very much on the skill of the vendor as a speaker and salesman. The travelling medicine show persisted longer in the United States and many of the quack doctors were highly skilled at selling. An account of the Tiger Fat and Vital Spark lecture given by Mrs Violet Blossom (or Madame V. Pasteur as she styled herself) is given in Chapter 4.

Aphrodisiacs were rarely, if ever, referred to as such, but more frequently appeared under the guise of restoratives or rejuvenators or simply tonics, possibly to avoid offending the sensibilities of potential customers. Nevertheless, when advertisements for such remedies appear, it does not take any great effort of imagination to appreciate precisely what is being promoted. Although personal salesmanship has had a long history in the advertising of patent medicines, it was the growth of the press in the eighteenth century that provided manufacturers with their most profitable means of access to a wider public. Most newspapers of the period, as now, relied heavily on the income from advertising to cover the costs of production and distribution which could not be covered by the selling price of the newspaper. Over recent years advertising has become a major industry in its own right and the costs involved in the promotion and marketing of a product can, in some cases, represent a significant proportion of its final retail price.

Although news items and pictures of naked girls tend to dominate the popular newspapers of today, at one time it was the advertisement sections that provided the most interesting and informative reading.

A very large proportion of the advertisements appearing in the days before legislation restricted their content promised patent medicines of astonishingly wide efficacy and potency — in other words what we would now call quack remedies. Only a fraction of this type of advertising remains in the national and local press, and it is to the less savoury end of the adult and underground press that we must turn to find the modern quack doctors.

Some idea of the extent of aphrodisiac advertising that was appearing in the newspapers of the eighteenth century can be gained from the results of a survey of newspapers in the City of Bath at this time.[3] Out of a total of 7988 advertisements covering 302 different medicines, 1651 were general remedies for debility, 534 were for the treatment of venereal disease (a particular problem in Bath at the time), and 301 were for the 'complaints of females'.

Advertisements of this era were usually of the 'puffing' variety; that is, they merely made exaggerated claims for a product. A few played upon the fears and gullibility of the readers by telling them that they were suffering from major disorders that could only be treated by the medicine in question. Assuming that a sufficient number of hypochondriacs would respond to these advertisements, success was assured. A particular feature of all these advertisements was the grandiloquent style of language that was employed. What the copy lacked in scientific accuracy was amply compensated for in literacy hyperbole. A particularly good example is provided by Dr Brodum's Nervous Cordial and Botanical Syrup. A sample of Dr Brodum's style appeared in *The Times* of 3 October 1798. The headline news of the day was Admiral Nelson's victory at the battle of the Nile, but at this time, and indeed until fairly recently, the front and back pages of *The Times* were entirely devoted to personal notices and advertisements. Turning to the back page we can read the following:

'TO THE PUBLIC. The increasing sale of Dr BRODUM'S NERVOUS CORDIAL and BOTANICAL SYRUP sufficiently evinces their efficacy in the cure of those complaints for which they are recommended. To the afflicted with Nervous Disorders, to those suffering from change of heat or climate, and to those who labour under weakness and relaxations, originating in a variety of causes, this medicine is recommended as a powerful restorative; and to thousands of young people growing old before their time, by having unguardedly plunged themselves in the commission of a solitary and deluding vice, they have happily been the means of recovery to the mind as well as the body, and exalted them from a state of melancholy and despair to one of health, peace, and happiness. This delusive habit, here alluded to, is not confined to the gay, the giddy, and the vain, for alas! the rich, the poor, the young, and those of riper years, nay even those of a serious and religious disposition are often drawn, by an unaccountable infatuation to the commission of this melancholy practice. All such should seriously attend to the observations and cases described in Dr Brodum's Guide to Old Age, price 5s., the eleventh edition of which is this day published. In that mirror they will behold the dreadful consequences they are producing to themselves, and may then perhaps be persuaded to retire from that road, which though strewed with flowers, is sure to lead them only to destruction.

The abominable practice that is the present subject of reprobation, produces those disorders in age that lurk unseen in the constitution until they have derived strength to occasion the most destructive effects, and when these baneful evils make their appearance, the valetudinarian is at a loss to ascertain the cause of his maladies.

The Botanical Syrup and Restorative Nervous Cordial may be had at the Doctor's, No. 9, Albion Street, in bottles of £1 2s., 11s 6d., and 5s 5d., duty included, and of Mr Robinson, Tunbridgeware manufacturer to their Royal Highnesses the Princess of Wales and Duchess of York, next to York House, Piccadilly; E. Newberry, corner of St. Paul's Churchyard; Ward, No. 294 Holborn, opposite Gray's Inn [here follow a dozen or so provincial agencies]. It may also be had by giving orders to the Newsmen.'

It is a sad consequence of the present permissiveness in society that such marvellously evocative euphemisms such as 'a solitary and deluding vice' or the better known 'love that dare not speak its name' are unnecessary and we can refer less tastefully but unambiguously to masturbation and homosexuality. The sample of advertising just given illustrates all the key features of this form of promotion, namely:

(1) Aim your product at the widest section of the community.
(2) Make the symptoms of the disorder as general and vague as possible.
(3) Imply that serious risks might be incurred unless treatment is undertaken without delay.
(4) Price the treatment on the basis that the level of expectation of

the customer is directly related to the price he has paid for the medicine.

The North American approach tended to prefer the blunt and un-compromising statement rather than the long-winded sermonizing of Dr Brodum. For example, in Cincinnati, Dr Raphael wrote of his Cordial Invigorant: 'The cordial stimulates the sluggish animal powers to healthy action and with vigor comes NATURAL DESIRE. It brings the system up to the virile point and KEEPS IT THERE.' Dr Raphael, like many of the entrepreneurs in his adopted country, had little time for subtlety in his advertisements.

The Testimonial

The testimonial has always been a valuable advertising ploy. Originally, an unsolicited testimonial from a gentleman of the cloth was considered to be the ultimate accolade, but as the influence of the church declined in society, the medical profession became the target for such soliciting. A letter from a physician to a Dr Taylor and published in *The Times* at the beginning of the last century gives an idea of the sort of thing that was written:

'Sir,
With the blessing of God having received a great cure in a scrofulous case, by taking your Scake's Patent Pills, I deem it incumbent on me to give you an account of it, begging that, for the good of others, you would make it as public as possible.'

The letter proceeds at great length to describe the case and the for-tuitous employment of the Patent Pills. No doubt many testimonials were completely authentic and written with the best possible motives and no thought of personal gain, but the scope for fraud is obviously great. Consequently, testimonials have tended to attract suspicion, and considerable care is taken nowadays to ensure that copies of original letters are lodged with the publisher of the magazine or newspaper in which the advertisements appear.

The most blatant example of a company actively soliciting for testimonials occurred in the case of Dr Gardner's Pink Tablets in 1907. The company offered a £100 prize for the best testimonial submitted, together with an unlimited number of runner-up prizes of one guinea for any other testimonials that they considered to be of sufficient merit to employ in their advertisements. Each example submitted had to be accompanied, of course, by a coupon from a box of the tablets. Customers were also not discouraged from sending in more than one entry for the competition.

Today, only marginally more subtle means are employed to obtain good original ideas for advertising slogans from the public. Competitions that involve a token element of skill and judgement appear regularly in magazines, and invariably include, as a tie-breaker, a sentence extolling the virtues of the product, to be completed within a given number of words. As with the infamous Pink Tablets, some evidence of purchase is also required with every entry. Recently, testimonials have tended to evolve into the phenomena of sponsorship and endorsement of products by celebrities. However, there is much old-fashioned reticence concerning the nature and the business of the sponsor. There has, for example, been much criticism of tobacco- and drink-company sponsorship of sporting events since neither product is associated with good health and fitness. In this climate it is highly unlikely that any product with even remote aphrodisiac or sexual overtones will be linked to an individual celebrity or sporting event. The nearest we have come is the well-known association of a former heavyweight boxer and a world motor-cycle racing champion with a brand of perfumes and deodorants for men.

Advertising Legislation

Good taste, self-esteem, fear and embarrassment have tended to discourage the numerous sufferers from debility or lack of manhood from admitting their deficiency in testimonials. On the other hand, the fear and embarrassment involved in sexual disorders in general have been of enormous value to the purveyors of restoratives, invigorants, tonics and remedies for the 'secret diseases' of both men and women. In order to overcome this entirely natural and understandable shyness, the quacks offered monographs on the self-diagnosis and cure of such disorders. Diagnosis by correspondence was also common and involved either the filling in of a questionnaire (in complete confidence of course), or the sending of samples of hair or urine for a complete diagnosis of the client's state of health. Not surprisingly, very few if any clients received a clean bill of health by this means.

Initially, it was the medical profession who agitated most strongly for legislation against the quacks, since they recognized that here was a significant source of practical competition for patients and the income that ensued. Not all physicians were concerned purely with protecting their self-interests, and the chief pioneer of legislation against ineffective and impure drugs in the USA was Howard Wiley. His efforts were largely responsible for the introduction of the Pure Food and Drug Act of 1906 and the founding of the Food and Drug Administration (the FDA, who still supervise the testing and licensing of new drugs in the USA). The later Food, Drug, and Cosmetic Act of 1938 effectively inhibited the promotion and sale of most patent medicines in the country in which they had been most prolific. Nevertheless, it is still possible to find advertisements in adult magazines for 'Pseudo-Hard-On' pills, 'Jungle Love' and 'Vice Spice'.

By the end of the nineteenth century, one of the most practical means of attacking the quacks, made possible by the rapid development of chemical techniques, was to analyse their secret remedies and potions and, more significantly perhaps, calculate the cost of the ingredients. To take just one example, Dr Lecoy's 'Vigoroids' were sugar-coated tablets containing: ferrous phosphate, quinine, strychnine, vegetable extract, sugar, starch and talc. The largest single ingredient was sugar, whereas the strychnine, which is highly poisonous at large doses, was only present at 0.005 grain (0.3 mg) and would only be likely to contribute a bitter taste to the pill at this concentration. The disorders for which this particular remedy was indicated are fairly evident from its name. (Further recipes of quack preparations are given in Chapter 4.)

The problem of misleading or extravagant advertising claims was not limited to the United States and Britain; in France, Germany, Italy and beyond, there were legions of young men to be persuaded of their nervous debility and offered immediate and expensive relief by the quack doctors. The very wording of their advertisements provides an insight into the nature and habits of the potential customer: at the same time as Dr Lecoy's Vigoroids were restoring the lost youth and vitality of Europe, in Calcutta, Kaviraj Keshab Lall Roy's Wonderful Remedies were fulfilling the same purpose. This particular charlatan had no less than four remedies specific for debility including *Ratiballav Moduck*:

'. . . a powerful aphrodisiac, which gives vigour to the nerves and organs; enhances the manly power, strengthens the nerves, and nourishes the blood vessels and muscles. It makes life cheerful and a blessing. It is what every husband should have and what every wife desires in her heart of hearts.'

A box of this precious elixir, sufficient for 15 days, was sold for 1 rupee. The other weapons against debility in Kaviraj Roy's armamentarium included *Chyavanpras*: '. . . radically cures nervous debility, general prostration, languor, and lassitude arising from excessive mental exertion and abuse of physical strength'. Alternatively, one could take *Jogeshwar Ghrita*: '. . . the best nerve tonic. An infallible remedy for nervous debility, loss of manhood, and diseases arising from over-exercise of the brain, excessive drinking, and sedentary habits'. And last, but by no means least, the Vigor Pill or 'Friend of the Deluded Youth' as he calls it. Kaviraj Roy, back in 1910, was clearly a talented and experienced operator who knew well the fears and foibles of his customers.

Decent, Honest, Legal and Truthful – Advertising Codes of Practice

In Britain, the Advertising Standards Authority was set up as a non-profit-making company to monitor and regulate advertising by means of a standard code of practice. Many products are covered by the Medicines Act 1968 and the Labelling of Food Regulations 1970, but there are still loopholes through which dubious remedies can be promoted. Since medicines as such are very strictly controlled in that they must be shown to have been thoroughly tested and then approved by the Committee on the Safety of Medicines, some entrepreneurs have described their products as dietary supplements or even vitamins. The prime example of this is the marketing of Gerovital or vitamin H (see page 87). The word 'vitamin', although scientifically precise, can still be used in a wider sense in advertising. Some vitamins have extremely dubious parentage indeed, and vitamin H is merely another name for the drug procaine, which acts principally as a local anaesthetic.

The Proprietary Association of Great Britain (PAGB) represents the manufacturers of medicines that are available to the public without prescription. They published their first Code of Standards of Advertising Practice in 1936, and it has since been regularly updated. All copy for advertisements or promotional material has to be submitted to the PAGB prior to its use. In the UK there is a mixture of statutory legislation (the Trade Description Act and the Retail Trading Acts) and voluntary self-regulatory controls. In addition, advertising of medical products has to satisfy the provisions of the Medicines Act of 1968.

It is an offence under the Medicines Act to issue false or misleading advertisements for medicinal products of any description. It is also an offence to recommend the use of a medicine for any purpose other than that specified by its original product licence. These restrictions have had negligible influence on the purveyors of sex aids and aphrodisiacs although, together with the British Code of Advertising Practice, they ensure that such products are only promoted in the more unsavoury adult magazines and not in the popular press.

The British Code of Advertising Practice includes examples of what are considered to be unacceptable claims or wording of advertisements. The Code states that advertisements should not misuse scientific terms or statistics in order to make a claim appear to have a scientific basis which it does not possess. Also, they should not play on the emotion of fear or exploit the superstitious. These strictures apply to products in general and there are more specific requirements which apply to medicines, treatments and appliances. For example, it is not permitted to refer to 'colleges', 'hospitals', 'clinics', 'institutes' or 'laboratories' unless they are bona fide establishments under the supervision of suitably qualified members of staff. Also, no advertisement should claim or imply the cure of any ailment, illness or disease, as distinct from the relief of its symptoms. (This is where the wording has to be very carefully phrased, and some quite ingenious means have been used to overcome this restriction.) No advertisement should encourage the indiscriminate, unnecessary or excessive use of products. This restriction is intended to overcome the fallacious but widely held belief that taking larger than necessary doses of a drug will increase the therapeutic effect. In practice, the taking of excessive doses merely results in an increased likelihood of unwanted and possibly unpleasant side-effects.

No advertisement should offer any product for a condition which needs the attention of a registered medical or other qualified practitioner. Sexual problems obviously fall into this category, but from the number of letters received by the editors of agony columns in popular magazines, many people with problems still find it impossible to discuss them with their own doctor and prefer diagnosis and treatment by post. Part of the code deals specifically with aphrodisiacs as follows:

'Advertisements should not suggest or imply that any product, medicine or treatment offered therein will promote sexual virility or be effective in treating sexual weakness, or habits associated with sexual excess or indulgence, or any ailment, illness or disease associated with such habits.'

Figure 1.1: The Covers of a Trade Pamphlet Advertising 'Mormon Elders Damiana Wafers' with Unmistakable Sexual Overtones. A more recent example of aphrodisiac advertising is illustrated in Figure 4.3.

In other words, you cannot advertise aphrodisiacs. Nevertheless, such advertisements do appear for promoting products which to all intents and purposes are aphrodisiacs. Examples of advertisements are shown in Figures 1.1 and 4.3.

2 THE ANCIENT TRADITIONS

Most of our knowledge has come down through written traditions which have been preserved in specific cultures and occasionally passed on to other cultures through trade and other contacts. In this way, for example, it is possible to trace a history of medical knowledge from the Egyptians to the Arabs and Greeks and thereby to the Romans. During the Dark Ages and until the Renaissance in Europe, much knowledge was considered heretical and was suppressed. The uneducated masses had to rely on folk knowledge, which has proved to be surprisingly resilient. The growth of travel and exploration which followed the Renaissance introduced knowledge from new cultures to Europe, notably from India and China. More recently though, anthropological surveys have found some quite small isolated tribal groups with their own body of medical knowledge, however primitive. There is always a large magical component in the application of this knowledge but common factors have often emerged between races and cultures in the use of plants, drugs and ritual ceremonies in the pursuit of sexual success and fertility. This chapter provides some accounts of the use of drugs and magic as aphrodisiacs in ancient times in order to indicate how little we have changed in our basic attitudes to sex. That there is nothing new under the sun is amply illustrated by the early reports of the use of opium and hashish. Even the love charms and potions of historic times will be recognized as still existing under a thin gloss of civilization. Over 5000 years of cultural development have not noticeably diminished the naivety and gullibility of man.

The Hindu Tradition

There are several classical works in Sanskrit literature on the subject of love, written mostly by poets and learned men of wide interests. The best-known surviving manuscripts are the *Anunga Runga*, composed by the poet Kullianmull some time in the fifteenth century, which concentrates on the different classes of men and women and their characteristics, and the earlier *Kama Sutra* of Vatsyayana, written some time between the first and fourth centuries AD. This last was one of the books translated into English by Sir Richard Burton and published privately in London in 1883 by the Kama Shastra Society. It can hardly be described as a titillating text and yet it has achieved a considerable reputation as one of the great erotic treatises of the East. It is still regularly reprinted and has been widely read in Europe and America in the hopeful belief that some of the famed mystical secrets of the Orient might be gleaned from its pages. In fact it is less an encyclopaedia on sex technique than a textbook on the correct social conduct in the caste-conscious society of India fourteen centuries ago. Although the society for which the *Kama Sutra* was written, and whose members sought Kama through sexual pleasure have long since passed away, some of the social etiquette described in the book would not be out of place in contemporary society.

The Hindu culture clearly recognized bodily cleanliness to be as important as spiritual cleanliness, and in the wealthier classes the knowledge and use of scents and aromatic oils were extensive. Perfumery was an art as well as a profession in those days. The short chapter in the *Kama Sutra* which deals with the methods for 'attracting others to oneself' includes a nice mixture of magic ritual, medicines of greater or lesser potency, and practical remedies and advice, most of which reappear in modern erotic writings and books on sexual technique. Some of the recipes contain local plant or animal ingredients whose Sanskrit names have not enabled them to be identified with certainty, so there still remains some element of mystery concerning Hindu aphrodisiacs.

Much of Vatsyayana's advice is eminently sensible, involving the use of cosmetics or perfumes to increase one's attractiveness, and the subtle arts of music and conversation to induce the appropriate mood in the object of one's desire. The essential skills of seduction have changed little over the centuries.

On the other hand, a typical magic ritual consists of the following: sprouts of the *vajnasunhi* plant are cut into small pieces, dipped into a mixture of red arsenic and sulphur, and dried seven times (this may seem rather excessive, but it must be remembered that the number seven has strong magical significance). The dried mixture, which is now in the form of a powder, is burnt and the smoke is observed: if a golden moon is visible through the fumes, then the amatory experience will be successful. In fact, the burning of most metals and sulphur in combination will tend to produce an impressive quantity of smoke. In industrial towns where smelting is carried out on a large scale, the only opportunity the general populace ever has of observing the moon is through a red smoky haze. The moon itself, of course, can provide an ideal romantic background for nocturnal assignations, and gazing at a full moon hanging low in the sky can have the most remarkable effect on some people.

Alternatively, if one is seeking an anaphrodisiac effect, then the powder from the recipe given above mixed with monkey excrement, and thrown over a maiden, will ensure that she is not given in marriage to anyone else. This hardly seems a surprising outcome under the circumstances!

There are several methods described in the *Kama Sutra* whereby a man may increase his sexual vigour. Most of them include milk and honey, which have always been recognized as foods capable of providing stamina and energy. There are marked similarities between these recipes and those described by Greek and Roman writers. One good example can be prepared as follows: take the testicle of a goat or a ram, boil it in milk and sugar, and the resultant liquor, it is claimed, will increase sexual vigour. An equally effective dish can be made from rice and sparrows' eggs, boiled in milk, and then mixed with ghee or clarified butter and honey. Both these recipes combine the magic symbolism of the generative organs of animals with ingredients likely to produce an immediate source of energy for engaging in physical activity.

A relatively straightforward recipe for a drink reputed to taste like nectar consists of ghee, honey, sugar, and liquorice combined in equal quantities, and then mixed with milk and the juice of fennel. It would not be difficult to build this cocktail today and it is claimed to be provocative of sexual vigour; however, the author has not tried it.

Numerous other foods and sweetmeats are described in the *Kama Sutra*, which the keen reader can try for himself, but they are not likely to have anything more than a psychological effect upon the libido. The regular appearance of highly nutritious food substances, such as ghee, milk, eggs and honey in these recipes provides the essential clue to the rationale behind their use. In a society where the normal diet is likely to be deficient in either proteins or essential vitamins, the addition of a highly nutritious supplement would tend to increase general vigour and produce an improvement in the mental state. A deficient diet will, over the long term, cause feelings of lethargy and indifference and a loss of sexual appetite; an extremely nutritious 'aphrodisiac' under these circumstances would certainly stimulate a sexual appetite which had been artificially curtailed by a pre-existing dietary insufficiency. Aphrodisiacs of this type are unlikely to be of benefit to the more than adequately fed Westerner.

From the practicality of diet we can now return to the mysteries of magic: 'By eating the powder of the *Nelumbrium speciosum*, the blue lotus, and the *Mesna roxburghii*, with ghee and honey, a man becomes lovely in the eyes of others.' Or: 'If the bone of a peacock or of a hyena be covered with gold, and tied on the right hand, it makes a man lovely in the eyes of other people.'

Modern advertising may have evolved in terms of the subtlety of the message, but the product and the promotion are not entirely unrecognizable today. Items of jewellery or adornment could be rendered even more effective by an enchantment: a bead made from the seed of the jujube, or from the conch, could be empowered to increase sexual attraction by the appropriate incantation from the *Atharvana Veda*.

More directly acting sexual stimulants involve principles common to other cultures. A mixture of powdered white thorn-apple, long pepper, black pepper and honey, applied to the penis prior to coitus, is reputed to make the woman subject to the man's will. The thorn-apple powder will contain atropine and hyoscine, which are both potent alkaloids with the properties of inducing an initial excitement phase followed by sedation and hallucinations. These drugs would be readily absorbed through the mucous membranes of both the penis and vagina, and might therefore be expected to have significant behavioural effects in both partners. The pepper extracts have a counter-irritant or rubefacient action, and would produce a burning or itching sensation around the area of application. In the male this can increase blood flow into the penis and help in the development and maintenance of an erection. The effects of the peppers would also be quickly transmitted to the woman, who would probably feel a mild irritation and an urge to urinate. Irritation of the clitoris would certainly increase the sexual desire. The honey could act as a useful lubricant, rendering the penis a greater ease of access into the vagina. This recipe as a whole, therefore, has a completely rational basis for its action and might well prove effective in practice.

Less savoury compositions employing, among other things, the remains of kites that have died from natural causes are given equal credence, although there is no earthly reason why they should have the slightest beneficial effect. Some recipes include heavy metals and there have been several recently reported cases of poisoning among immigrants from the Indian sub-continent who have used traditional aphrodisiacs sent to them from home (see page 95). Nevertheless, the early Hindu masters of eroticism have left us a valuable legacy of culinary and olfactory delights aimed at encouraging and increasing sexual pleasure.

China and the Far East

Since the Chinese represent one of the oldest civilizations in the world, it is not surprising that they possess an elaborate and well-established herbal tradition. The pharmacopoeia of Emperor Shen Nung, which describes Indian hemp, the opium poppy and aconite among other equally potent preparations, was written about 3000 BC. Modern Chinese pharmacopoeiae rely heavily on these very early traditions. As with the Japanese, the isolation from Western influences has resulted in an independent and still flourishing trade in herbal remedies. Even today there is much resistance to the use of Western medicines in China, for social as well as economic reasons. Chinese herbalists abound in Hong Kong, where the standard of living is much higher than on the mainland and where modern drugs are readily available, so it would seem that the traditional approach survives because it is preferred by the patients rather than as an economic necessity.

This 'alternative' medicine maintains a fascinating emphasis on the manufacture and sale of substances to promote longevity in general and restore sexual vigour in particular. The early herbalists discovered the basic technique of infusing leaves in hot water in order to extract the active principles (this is still the practical means for making tea), and they also discovered the value of the ginseng root. Every herbalist's shop in Hong Kong has some of these humanoid roots on display, in addition to deer antlers and other less savoury parts of animals. The price of many of these items, which often possess a more than passing resemblance to the male or female human genitalia, can be very high. A 'tiger penis' would currently retail at about £100.

The *Chung Yao Chih* (*New Chinese Materia Medica*) has been re-printed in English in a set of three volumes between 1959 and 1961, and it contains an abundance of information. Unfortunately, like most other materiae medicae derived from early sources, it consists largely of a catalogue of plants with a list of their properties, but without any specific evidence to support the numerous claims made. It is not easy under these circumstances to separate the potentially useful from the totally inert recipe. Nevertheless, ginseng preparations have proved to be of value and are currently gaining widespread acceptance in Europe and America. The specific properties of ginseng in all its forms are discussed in some detail in Chapter 4, but the following list will give some idea of the range of other plants with reputed aphrodisiac properties:

Abutilon indicum
Adenophora polymorpha
Aquilaria malaccensis
Artocarpus communis
Balanophora involucrata
Bombax ceiba
Cinnamomum verum (the source of commercial cinnamon)
Curculigo orchiodes
Cyasthostemma micranthum
Cynomorium coccineum
Dryobanalops camphora (an aromatic herb and the source of camphor)
Durio zibethinus (the milky sap acts as a counter-irritant)
Euchresta horsfieldii
Eucommia ulmoides
Eupatorium fortunei (Joe Pye weed)
Hygrophila auriculata
Juniperus chinensis (the source of juniper berries, the main flavouring component in gin)
Magnolia officinalis
Mallotus poilanei
Mucuna gigantea
Myristica fragans (nutmeg)

Ophiopogon japonicus
Orobanche ammophila
Orophea setosa
Panax ginseng
Panicum sarmentosum (chewed by the Malays together with betel nut as a social aphrodisiac)
Photinia serrulata
Pimpinella alpina
Pistia stratiotes
Potentilla discolor (cinquefoil)
Prunus japonica (cherry)
Psoralea corylifolia
Scaevola taccada
Schisandra hanceana
Smilax myosotiflora
Strychnos nux-vomica (the source of strychnine and highly poisonous)
Styrax benzoin (snowbells; the source of the aromatic benzoin)
Talinum pateus
Thuja orientalis
Tribulus terrestris
Valeriana hardwickii (related to the European valerian)
Zanthoxylum piperitum

Most of the plants in this list have a variety of other applications in addition to their supposed aphrodisiac action. *Balanophora*, for example, is recommended for piles; the gum from *Bombax* acts as an astringent and can be used to staunch bleeding in the same way as a stick of alum. The dried stalks of *Cynomorium* are used to treat kidney trouble, impotence, sterility and constipation, and for strengthening the back. The oil from *Juniperus* contains the essential oils pinene, limonene, cadinene, sesquiterpene, cedrol and cedrene, as well as valerianic and decyclinic acids. Juniper oil is recommended in the treatment of gonorrhoea, ringworm and eczema. It is interesting to speculate whether gin, which contains extract of juniper, might share these valuable properties.

The fruit of *Schisandra* contains citric, malic and tartaric acids, carbohydrates, resins, vitamin C, manganese, iron, silicon, phosphorus and calcium, in addition to some alkaloids of unknown chemical structure present in the seeds. The fruit is unusual in that it is reputed to possess all of the five basic flavours, namely: sweet, sour, bitter, hot and salty. The tropical water plant *Pistia* has an extremely bitter taste, but the roots are cooked and used as an aphrodisiac by the natives of New Guinea. In China, the same plant is used externally for treating boils, swellings, contusions and syphilitic eruptions. The root of *Talinum* is used as a cheap substitute for ginseng (something of an expensive luxury among Chinese herbal remedies). The leaves of *Zanthoxylum* are recommended as an antidote to bites and strings; mixed with vinegar they will cure hiccups.

Chinese Prescriptions

There are several specific prescriptions for the cure of impotence, and the following recipes are typical, although their efficacy cannot be guaranteed:

Prepare a 300 c.c. decoction of one or more of the following:

Selinum monnieri, 8 g
Epimedium macranthum (the leaves contain the glycoside icariin and an unknown alkaloid), 7 g
Plantago major (this contains sugar, citrate, oxalate, emulsin, invertin and aucubin), 10 g

Extracts given to animals are reputed to increase their rate of copulation and increase the volume of seminal secretions.

Achyranthes bidentata, 7 g
Cimicifuga foetida (the root is very bitter but has little pharmacological activity), 7 g
Dioscorea sativa, 5 g
Acanthopanax spinosum (the roots contain essential oils, resins and vitamin A), 6 g
Asparagus lucidus, 10 g

The recommended dose is 100c.c. of the decoction taken three times daily.

As a tonic for improving the general vigour, one or more of the following can be recommended:

Artemisia vulgaris (mugwort; the leaves are bitter and contain cineol, thujone, tannin and artemisin), 7 g
Kochia scoparia, 7 g
Selinum monnieri, 7 g
Epimedium macranthum, 8 g
Sophora flavescens (the roots contain the alkaloid matrine and are astringent), 7 g

Plantago major, 10 g
Kochia scoparia (the Belvedere cypress), 12 g
Selinum monnieri (the bitter seed contains 1.3 per cent essential oils, in particular borneol, pinene, camphene and terpineol; it is pre-scribed as a specific stimulant and aphrodisiac, and a cure for piles), 12 g
Eugenia caryophyllata (clove tree; the floral buds, which are sold in this country as whole cloves, contain 15-19 per cent of volatile oil, of which about 85 per cent is eugenol; it is antiseptic, antispasmodic, and a general stimulant), 10 g

The recommended dose of this mixture is 5 g, presumably to be taken orally. Together with the other recipes listed, it can no doubt be obtained from a reputable Chinese herbalist today — at a price.

Although the Chinese represent one of the earliest and most sophisti-cated cultures, the use of aphrodisiacs was not limited to the most advanced civilizations even though their written records tend to be more complete. Even the nomadic tribes had their own traditions for making aphrodisiacs, and the French writer Foucher d'Obsonville has vividly described a practice employed by the Tartars, who believed firmly in the effectiveness of animal genitalia for this purpose. The stallion was the source of the most potent material, obtained as follows:

'Les palfreniers amènent un cheval de sept à huit ans, mais nerveux, bien nourri et en bon état. On lui présente une jument comme pourre la saillir, et cependant on le retient de façon à bien irriter ses idées. Enfin, dans le moment ou il semble qu'il va lui être libre de s'élancer dessus, l'on fait adroitement passer la verge dans un cordon dont le noeud coulant est rapproche au ventre, ensuite, saisissant à l'instant où l'animal paraît dans sa plus forte érection, deux hommes qui tiennent les extrémités du cordon le tirent avec force et, sur le champ, le membre est séparé du corps au dessus du noeud coulant. Par ce moyen, les esprits sont retenues et fixes dans cette partie laquelle reste gouffrée; aussitôt on la lave et la fait cuire avec divers aromatiques et épiceries aphrodisiaque.'

This rather unpleasant operation is carried out in the belief that the power of the male genitalia lies in the penis rather than the testicles. This somewhat primitive view persists in the belief in the efficacy of natural products which merely resemble the penis. To most people, the fully engorged penis of a rampant stallion was not a common feature of their life, and this Tartar practice is probably confined to cultures in which the horse still features prominently as a means of transport and a symbol of wealth and social standing.

The Near East

That the Babylonians and Assyrians practised magic is clear from the information derived from translations of the early cuneiform tablets found by archaeologists at various sites in Arabia. A large part of the knowledge and learning accumulated by later Arabic scientists was also probably derived, at least in part, from the traditions of earlier cultures established in the same geographical areas.

Much of the early magic dealt with methods of protection against malign spirits, and the means for expelling demons from the sick, but some specific love charms have been identified. For example, the brain of the hoopoe mixed into a cake, or a lamp wick inscribed with the appropriate invocations and burnt, was being used over 3000 years ago. The bones of a frog, buried for seven days and then exhumed, could be employed to detect whether one was an object of the love or hate of some other individual: if the bones floated after being thrown into water, it showed love, but if they sank, then hate was indicated.

The Egyptians were also great magicians who believed that all knowledge, both natural and spiritual, was derived directly from the gods. The herbalists in ancient Egypt served also as priests, and their remedies invariably involved an element of magic. There is no doubt that they were aware of the beneficial properties of many plants which are recognizable today, but they regarded these properties as evidence of the existence of the spirit of the god within the plant rather than being due to some active substance present in the plant. A certain degree of self-interest by the priestly classes is also evident here, for if a plant were to be equally effective in the hands of a layman, the power and influence of the priests would be severely diminished. It was therefore essential to maintain a belief in the necessity for the appropriate magic rites and incantations to be carried out in order that a herb should achieve an effective cure.

Drugs used directly as aphrodisiacs are not in evidence from the accounts of medical practice at the time, but amulets used as charms to ensure fertility have been found in tombs, and they obviously played an important role in protecting the wearer from evil influences. More details of love charms have come from Hebrew manuscripts: the Jews learnt much magic during their period of captivity in Egypt, and there are numerous references in the books of the Old Testament to the mystical practices of the time. A Hebrew manuscript found at Mossoul, and quoted by Thompson, includes the following examples:

'Write and put into the fire, Alp, Sulb, Nin, W'Alkom, Apsaka, Bal in the heart of —— daughter of —— for love of —— son of —— like the love of Sarah in the eyes of Abraham.'

Similarly,

'Thou shalt fashion parchment after the fashion of male and female and on the picture of the female write, Bla Bla Lhb Lhb Lhb Hbl Hbl Hbl, and on the other write, Zkr Zkr Zkr Rhz Rhz Rhz Krz, and thou shalt put them together, front and back, and thou shalt put them in the fire.'

Alternatively,

'To bring a disdainful woman, let him write on one of her garments and make a wick of it and burn it in a pottery lamp, this, Halosin Halosin Alosin Alosin Alosin Sru'in Sru'in in that ye come and assemble in the body of — daughter of — and harass her that she eat not, until she come near me and do the pleasure of me, — of —'

The principles of sympathetic magic that dictated the form of charms such as these are considered in more detail in Chapter 3.

The Greek Tradition

Like the Arabs, the Greeks also derived a considerable body of knowledge of magic from the writings of the earlier civilizations that flourished around the Mediterranean. The magic love charms of the sorceresses Medea and Circe have survived in the stories from Greek mythology which can still be read today. A particular charm, mentioned by Aristophanes and Lucian, consisted of the ritual of 'drawing down the moon' (see Figure 2.1). The ceremony involves the invocation of the goddess Hecate and her attendant ghosts. A clay figure is made which is imbued with the power to fetch the subject of one's desire, who will then fall under one's influence. According to Thompson (writing in the 1920s), this ritual is still practised in Greece today. Alternative means of inducing love involved the making of wax images, a practice which has survived in European witchcraft. In one piece of magic described by Thompson, the man is directed to make the figure of a dog in wax mixed with gum and pitch, eight fingers long, and write certain 'words of power' over the region of the ribs. A tablet is then to be inscribed with 'words of power' and on this the figure of the dog is placed, and the whole put upon a tripod. The man then has to recite the 'words of power' written on the dog's side and the names on the tablet. Should the dog then snarl or snap, the lover will not gain the object of his affections, but if it barks, she will come to him.

In a second method the lover makes two wax figures, one in the form of Ares (the Greek god of war), and the other of the woman. The latter is made in a kneeling position, with her hands tied behind her back and the figure of Ares placed standing over her with his sword at her throat. The names of demons are inscribed on the limbs of the woman, and then thirteen bronze needles are inserted into the limbs. The man recites the words, 'I pierce the [mentioning the limb] that she may think of me'. Other words are written on a metal plate, which is tied to the figure with a string containing 365 knots. Next, the entire device is buried in the grave of someone who has died young, or by violent means. Finally, an incantation is recited to the gods and the lover will thereby obtain his desire.

This bizarre example of a kind of sympathetic acupuncture, associated with complex conjurations and a petitioning of the gods, is only one instance of an irrational human activity found throughout the world among tribes and cultures at all levels of development. (Even contemporary educated man is not immune to the activities of the

Figure 2.1: The Ceremony of Drawing Down the Moon. From an illustration on a Greek vase of about 200 BC. (Based on a drawing in *Mysteries and Secrets of Magic* by C.J.S. Thompson)

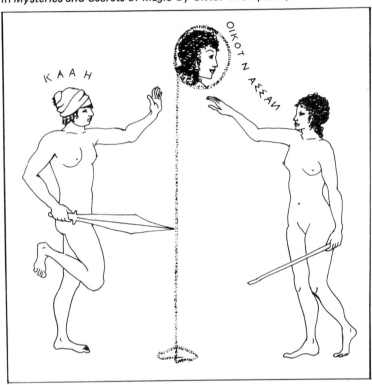

plausible thaumaturgists, a fact which will become apparent in the next chapter.)

In the example just given, the magical significance and thus the value, for use in rituals of this type, of the body of someone who has died violently, generated a lucrative trade in corpses long before they became a necessity for unscrupulous anatomists. At the height of the ascendancy of the Roman Empire it was the bodies of gladiators who had died in mortal combat in the amphitheatres of Rome and other cities which attracted the best prices. In fact, the Romans inherited most of their knowledge of magic, herbs, and medicine from the Greeks.

The Roman Tradition

The pre-Christian Roman Empire has never been considered as providing an exactly ideal example of a society with high moral standards, although the basis of Roman Law did emerge in principle, if not always in practice. We now tend to think of the period of the emperors as a time of violence and sensual excess, and the orgy is still regarded as a peculiarly Roman event. Caligula and Nero in particular achieved lasting reputations for promoting sustained sexual debauchery at a level that has probably not been equalled since. It is therefore to be expected that an age of widespread moral laxity would provide a large number of recipes for love potions, philtres and aphrodisiacs of all types.

Ovid, in his *Ars Amatoria*, which was written during this period of decadence, provides a very thorough and surprisingly relevant treatise on the techniques of seduction, although it is hard to imagine why the citizens of Rome at that time needed such a textbook. Instructions for the preparation and use of aphrodisiacs can be readily gleaned from Pliny's *Natural History*, in which the concoctions he describes range from the merely unlikely to the downright disgusting. A strong element of magic pervades many of these recipes, whereas others are simply herbal mixtures. Some of the more interesting 'facts' from Pliny's writings are set out below.

Rocket (*Brassica eruca*): 'a great provocative of lust'.

Wild parsnip: 'Orpheus states that it is an aphrodisiac, the wild variety being more potent than the cultivated and especially when it has been growing on strong soil. It also cures toothache, scorpion stings, hysterical suffocations, bowel pains in women, dysentery, dropsy, strangury, tetany, pleurisy, epilepsy and indigestion.' With such an obviously efficaceous vegetable growing wild, one is tempted to wonder what medical science has been missing for the last 2000 years.

Fortunately for us, Pliny lived in an age unfettered by scientific rationalism, educated scepticism, or a code of advertising practice. Very few of his recommendations are likely to have any specific action, and many have been derived wholesale from earlier Greek writers. For example: mallow: 'attached to the arms of patients with spermatorrhoea it provokes lustfulness'. According to Xenocrates, a physician who lived in Cilicia at the time of the emperor Tiberias: 'The seed sprinkled on the genitals will increase sexual desire in males to an infinite degree'.

Xenocrates clearly did not believe in understatement. Among other common garden plants, the carrot and the houseleek were also held in great esteem. That the Greek word for the carrot was 'philtron' is immediately suggestive of the role it may have played in love potions. We should now return to Pliny, however, who writes of nettle (*Urtica dioica*) as follows: 'When animals refuse to couple, it is recommended to rub the sexual organs with nettles'. It has been supposed that the Romans introduced the common nettle into Britain as a source of herbal remedies (nettle beer, nettle tea, etc.) and as a cure for the common cold, which they also brought; but mutual flagellation for sexual purposes is certainly an alternative possibility that should not be overlooked. The fine hairs on the nettle leaves contain a high concentration of acetylcholine as well as histamine, and both these substances contribute to the burning and itching rash which the leaves produce. This rather potent rubefacient action could serve a useful purpose in sexual stimulation. A sufficiently masochistic experimenter might find in nettles a practical means of encouraging an otherwise unresponsive male member into action. A similar but far more dangerous application of this counter-irritant technique provides the rationale for the use of Spanish flies or cantharides as an aphrodisiac (see Chapter 5).

Pliny also recommends another readily available plant, *Scandix* (the chervil), which 'reinvigorates the body when exhausted by sexual excesses and acts as a stimulant upon the enfeebled powers of old age'. This plant was clearly an invaluable aid for the ageing debauchees of ancient Rome. *Solymos* or *limonia* (possibly a species of thistle) was claimed (at least by Hesiod and Alceus) to be aphrodisiac when taken in wine. These authors are further quoted by Pliny as follows:

'While it is in blossom, the song of the grasshopper is louder than at other times, women are inflamed with desire, and men less inclined to amorous intercourse; it is by a kind of foresight on the part of nature that this powerful stimulant is then in its greatest perfection.'

The root too, after removal of the pith, has a valuable application: it 'corrects the noisome odour of the armpits'. Occasionally, however, even Pliny found his credulity stretched to the limit by some of the legends he was recording. For example, he quotes Threophrastus as describing a certain plant which, by its mere contact, has enabled a man to complete the sexual congress as many as 70 times! Unfortunately, Pliny has omitted to mention the name of this remarkable plant.

Other recommended herbs include *Xiphion* (*Gladiolus*) and *Cynosorchis* (an orchid), whose root tubers are supposed to resemble dogs' testicles. In fact, the word 'orchid' is derived from the Greek for testicle. Also mentioned are asphodel, erythraicon (*Erythronium dens canis*), and abrotonum, which is related to wormwood. Pliny is not, however, completely indiscriminate in his recommendations: *catomace* (which is possibly *Ornichopus compressus*) he dismisses as: 'a Thessalian plant which it would be a mere loss of time to describe seeing that it is only used as an ingredient in philtres'.

The Romans also found use for a number of anaphrodisiacs, including Caecilian lettuce, known also as *astytis* or *eunychion*:[6] 'it having the effect in a remarkable degree of quenching the amorous propensities'. Other anaphrodisiacs were nasturtium, wild mint and purslain (*Euphorbia peptis*), which has the additional properties of promoting menstrual discharge and the evacuation of intestinal worms. In fact, all species of *Euphorbia* are more or less drastic purgatives, and are quite likely to have a detrimental effect on any amorous inclinations.

Proceeding to some of Pliny's less savoury suggestions contained in his 'Marvellous Facts Concerned with Animals', we find a rich treasure of unpleasant marvels:

'The right side of an elephant trunk attached to the body with the red earth of Lemnos acts powerfully as an aphrodisiac. However, the skin of the left side of the forehead of a hippopotamus attached to the groin acts as an antaphrodisiac.'

This latter observation seems to be entirely credible under the circumstances.

'The right lobe of a vulture's lung attached to the body in the skin of a crane acts as a powerful stimulant in males. A lizard drowned in man's urine has an antaphrodisiac effect upon the person whose urine it is, also snail excrement and pigeons' dung taken with oil and urine. Similarly efficaceous are the testes or blood of a dunghill cock placed beneath the bed.'

Also:

'Hairs from the tail of a she-mule taken while she is being covered by the stallion will make a woman conceive against her will if knotted together at the moment of congress.'

This is of course a purely magic charm relying on the sympathetic magic of the knot-tying as a symbol for sexual union. The stallion also acts as a token of sexual power. Another magic rite (which even Pliny found dubious) consisted of taking the ashes of a spotted lizard and wrapping them in linen. Held in the left hand they act as an aphrodisiac, and transferred to the right they will have the opposite effect. There is a stereospecificity in many magic charms which is highly significant. One last example, which would undoubtedly be successful under any circumstances, consists of: 'the dust in which a she-mule has wallowed [which], sprinkled upon the body, will allay the flames of desire'.

Whether people actually used any of these recipes or charms for influencing sexual desire in either a positive or negative way must be open to doubt, but there is no doubt that the very extensive use of love philtres in ancient Rome led, at one stage, to the introduction of strict laws preventing their sale. Philtres and potions were a popular means of administering poison to enemies, and it was probably this fact rather than any moral reason which was responsible for their banning.

The Arab Tradition

The Arab civilization, which reached its peak around Baghdad in the thirteenth century, produced enormous advances in medicine, chemistry, astronomy and philosophy. This knowledge permeated slowly into Western Europe, brought back by crusaders returning from the Holy Wars. It was the Arabs who kept alive the medical knowledge of Galen at a time when Europe was undergoing a period of ignorance, superstition and religious intolerance. Many of the Arab medical texts have survived, and among these there exist specific aphrodisiacal books containing advice on sexual matters, prescriptions for sexual disorders and, of course, lists of aphrodisiac drugs and recipes. The best known of all these books is undoubtedly *The Perfumed Garden* of the sixteenth-century writer Shaykh Umar ibn Muhammed al-Nefzawi, translated by Sir Richard Burton. The book was originally published under the auspices of the Kama Shastra Society of London of which Sir Richard Burton was a member. The society existed to translate and publish erotic oriental works, and had already privately issued translations of the *Kama Sutra* and the *Ananga Ranga*.

The Perfumed Garden is a relatively modern work and much earlier references to prescriptions for sexual problems can be found, for example, in the formulary of Al-Khindi writing in the ninth century. An even earlier manuscript attributed to Galen himself contains two treatises, 'On the Secrets of Women' and 'On the Secrets of Men'. These were apparently written for Queen Filanus and Qustas al-Qahraman, respectively. Queen Filanus wished to know the means by which she could improve her bodily condition and restore it to health and beauty, and also the means by which she could inflict difficulties and distress upon the bodies of her enemies and rivals: in other words, she required a formulary of drugs to cure disease and drugs to induce disease. The recipes consequently appear in pairs, one to have a positive effect, the other to have a negative effect. We will now consider some examples.

'Drugs that excite the desire of women so that they go wandering around, leaving their homes, looking for sexual satisfaction, throwing themselves before men and searching for a good time: aged olive oil, orchis, garden carrot seed, turnip seed, ash of the leaf of the oleander, pulverized burnt nasturtium, dry alum, magpie excrement, powdered willow leaves and piths of fine dates.' A portion of each is ground up with coconut milk into tablets and dried. These are then pulverized and mixed into a syrup with juice extracted from roses. One tablet is taken for three days during the menstrual period. The recipe ends: 'It has the effect we described if God wills.' This is the let-out clause to account for any failure of the treatment.

For the opposite effect we find drugs that cause women to detest sexual intercourse in a way that they ignore it and forget about it: 'Forget-me-not, camphor, burnt goose bone, burnt paper, burnt wolf teeth, dried gall of a male goat, spikenard, yellow amber, caraway and dried white truffle.' One weight of each is taken and kneaded with rose oil and divided into three doses. These are applied by means of a tampon as follows: the first dose is applied while the woman sits in a bath of warm water for an hour. Two hours later it is removed and a second dose applied while the woman sits in a bath of cold water for an hour. Two hours later this is removed and the third dose is inserted and retained overnight. The authors gives no indication of how long this treatment might be expected to work. There is clearly a considerable element of magic in these recipes and this becomes clearer whenever more specific remedies are described:

'Drugs which make women detest lesbianism even if they madly lust for it: burnt copper, burnt vitriol, white vitriol, green vitriol, blue vitriol, glass, merlin, gallnut, pomegranate rind, marcasite and sweet hoof (this latter is thought to be the shell of a marine snail).' Two mithqals of each are taken and kneaded with rose fat obtained by using olive oil. The mixture should be carried in the vagina continuously for five days.

In contrast there are drugs that make lesbianism so desirable to women that they would keep busy with it and passionately lust for it forgetting all about their work: 'Narcissus bulb, yellow amber, wild rocket seed, wild rue seed, male fern, burnt dried cardamon, burnt bull's hoof, squill, galbanum, wild carrot seed, dill seed, serpolet seed, iron rust, burnt goat's hoof and blue lily.' This is administered in a similar fashion to the previous recipe, but in this case the author assures us that it will act for six months.

Other recipes for women include means of curing nymphomania and strengthening or softening the hymen. Although no one would now take any of these recipes seriously, they do provide an interesting insight into the sexual customs of the time. In 'On the Secrets of Men' there are some even more revealing specifics including 'drugs which increase men's desire for women so that they forget about boys and eunuchs', 'drugs which raise men's impatient desire for paederasty', and 'drugs which bring about addiction to masturbation'. The recipes are of a similar pattern to those already described and would be equally unpleasant.

In spite of the crude nature of these preparations, there is no doubt that the Arabs were accomplished chemists. They invented the technique of distillation and were able to use alcohol as a solvent for substances that would not dissolve in water. This was of particular value in the preparation of perfumes and essences from plants. All the erotic Arabic texts emphasize the value of perfumes, scents and cosmetics in increasing the pleasure obtainable through sex, and the very title *The Perfumed Garden* indicates the importance of this aspect of sensual delight.

The Perfumed Garden is probably the most widely known book of sex instruction. It was for many years banned from publication and only in relatively recent years has it become generally available as the social climate has become less inhibited. Although the major part of the book deals with sexual techniques, there are some recipes for improving sexual performance, some of which can be traced back to the original Galenic writings. The fruit of the mastic tree, pounded with oil and honey and drunk first thing in the morning, is claimed to increase the vigour. Another recipe consists of a glass of very thick honey, twenty almonds and a hundred pine nuts taken every night for three nights before retiring to bed. This relatively pleasant medicine is supplemented with powdered onion seed in honey. Less savoury is the fat from the hump of the camel, melted down and applied to the penis.

The Arabs were also aware of the stimulating effects of hot spices applied to the sexual organs as rubefacients. The author of *The Perfumed Garden* recommends macerated cubeb-pepper or cardamom grains or, alternatively, a mixture of pyrether, ginger and ointment of lilac. The most potent substance, however, appears to be chrysocolla, which is thought to be borax:

> 'You will likewise predispose yourself for cohabitation, sensibly increase the volume of your sperm, gain increased vigour for the action, and procure for yourself extraordinary erections, by eating of chrysocolla the size of a mustard-grain. The excitement resulting from the use of this nostrum is unparalleled, and all your qualifications for coitus will be increased.'

It is difficult to imagine how such a small quantity of a relatively innocuous chemical, which is a very feeble bacteriocide most commonly associated with gargles or mouthwashes, should have such profound and astonishing effects. If chrysocolla is indeed only borax, then the effects must be psychological.

There are also cures for the various forms of male impotence about which the author writes most knowledgeably. For overcoming premature ejaculation he suggests eating a stimulant pastry containing honey, ginger, pyrether, syrup of vinegar, hellebore, garlic, cinnamon, nutmeg, cardamoms, sparrow's tongues (possibly the seeds of the ash tree), Chinese cinnamon, long pepper and other spices. For temporary impotence he recommends a similar mixture of ground spices to be taken with water or honey.

One chapter in *The Perfumed Garden* is given over to 'prescriptions for increasing the dimensions of small members and for making them splendid'. All the prescriptions given would tend to increase the blood flow into the penis and perhaps help in gaining an erection, but there is no reason to believe that the effects might be permanent. The least dangerous consists of rubbing the member with warm water and then anointing it with honey and ginger which is to be rubbed in sedulously. The resultant erection is more likely to owe more to the physical manipulation than the nature of the anointing mixture. At the other extreme is the use of hot pitch, which would have been an extremely risky procedure. Bruised leeches and asses' members also feature in some most unlikely treatments which are reminiscent of the nostrums found in the writings of early Roman and Greek authors.

It would be a mistake to assume that this great armamentarium of aphrodisiac potions and remedies has entirely disappeared from use. A Dr Tacquin travelled widely in Morocco between 1910 and 1915 and wrote of his experiences in the *British Medical Journal*.[2] On the subject of aphrodisiacs he reported that premature impotence was very common in the native population, presumably as a result of the excesses to which polygamy leads, so that immoderate use was made of all the known aphrodisiacs and of the strongest condiments. Interestingly, the German government was active at the time in importing *Johimbine* (yohimbine) into Morocco where it was used widely. (The reputation of yohimbine as an effective aphrodisiac has only recently been discredited: see page 217.) Apart from the use of drugs and spices for this purpose, some magic charms were also employed. Dr Tacquin mentions one in which, in order to excite love in a woman, the man would give her to drink, without her knowledge, a sample of his semen mixed with sugar. In primitive parts of North Africa and the Near East it is quite likely that these unsavoury practices still persist today.

3 MAGIC CHARMS, POTIONS AND PHILTRES

Sex and Magic

Magic, as practised by societies throughout the world, is the art of attempting to influence the course of events by occult or secret powers, involving either the casting of spells by incantation or the manufacturing of charms. Even the recent rapid growth of science and technology has not eliminated all the elements of magic and superstition from our modern society. Many skyscrapers have no thirteenth floor, some airlines have no seat numbered thirteen, and cars painted green are relatively unpopular. Magic was the precursor of both science and religion, and its potency in primitive and some not-so-primitive societies should not be underestimated.

In the past, fertility was the all-pervading influence on life, whether in terms of human fertility or the fertility of animals and crops grown for food. The short life expectancy and high infant mortality rate necessitated a high birth rate in order to maintain a strong tribe or community. The fertility of the fields and the domestic animals meant the difference between survival and starvation. If Nature chose not to bestow this most precious gift upon a tribe, the result was almost certain death. As a result, much ancient magic sought to ingratiate Man with Nature and thereby ensure his survival. The magician or sorcerer in such a community therefore played an important role and was not merely a peripheral figure. We would now refer to the people who consulted these magicians or witches as pagans, in the sense of their being heathen or unenlightened, but it is worth remembering that the word pagan derives originally from the Latin, *pagani*, the country folk, and as such originally had no religious connotations.

Aspects of the pagan belief in fertility rites are still evident today in the form of the traditional maypole. Dancing round the maypole is essentially a fertility rite, and it is no coincidence that a young girl, usually a virgin, is chosen to be Queen of the May. She too is a symbol of potential fertility. An even more obvious example of an uncompromising symbol of fecundity is provided by the very evident phallus on the Cerne Giant, a large figure cut into the chalk of Trendle Hill in Dorset in England and still visible today. He probably functioned as a symbol of fertility for the surrounding farming community.

The fact that he has survived from the Iron Age, about 1000 BC, to the present day suggests that no intervening society has felt sufficiently confident in itself to let him revert to grassland.

The Principles of Magic

Before entering into detailed descriptions of love charms and potions, it is worth considering the principles upon which the practice of magic is based. These have been most clearly set out by J.G. Frazer in his classic work *The Golden Bough*, which he describes as a study in magic and religion. Frazer conveniently divides the principles of thought into two laws:

(1) *The Law of Similarity*. This proposes that like produces like, or that an effect will resemble its cause. Thus, a magician believes that he can bring about an effect by merely imitating it; for example, sticking pins into a wax image will injure the person on whom the image is based.

(2) *The Law of Contact*. This law states that things which have once been in contact with one another will continue to exert a mutual influence even after they have become physically separated. This contagious magic, as Frazer describes it, implies that parts of the body which have become separated from it (hair, nails, teeth and so on) can still influence the person of whom they were once part. In sexual magic it is the afterbirth, the menses and semen which are used to make charms or cast spells. Even items of clothing or personal property are endowed with the potential to influence the owner.

The most potent forms of sympathetic magic employ both principles to ensure their success, so images are made incorporating the victim's hair or nail parings. Not that all sympathetic magic was intended to be injurious: charms were often used for wholly beneficial purposes, although in the case of love charms they were occasionally used in an attempt to achieve success with an otherwise unwilling partner.

The Law of Similarity has given rise to the Doctrine of Signatures which the herbalists follow in order to discover specific remedies for disease. This theory and its consequences are discussed more fully in the next chapter. It explains, for example, why the orchid has a reputation as an aphrodisiac. The plant grows from a tuber which closely resembles the human testicle both in size and shape. The similarity is compounded by the fact that most orchids still retain the tuber from the previous year's growth so that a pair of tubers are found beneath each plant. Both the Law of Similarity and the Law of Contact have evolved into the science of homoeopathy. Although homoeopathic medicine is still widely practised and there is evidence that some patients are actually helped by it, the precepts upon which it is based have, at present, no scientific basis.

Charms and Taboos

Practical magic exists by a system of charms and taboos. Charms represent the positive side in that they are things you have to do in order to gain some particular purpose. Taboos are the things that you must specifically not do in order to avoid some particular consequence. In the simplest and most familiar example, good luck can be assured by crossing the fingers (a charm) and not walking under ladders (a taboo). A hole in a rock is a symbol of a woman's birth passage and so the act of passing through such a hole would increase fertility. There is a rock in Aberdeenshire where women still squeeze through the narrow passage known as the Devil's Needle in order to attract luck. They are probably not aware that they are re-enacting a primitive fertility rite. The symbolism of the vagina is much less common than the phallic symbol. Many lucky charms are strongly suggestive of the male member. Nowadays a single shark tooth on a necklace or, in America, the John the Conqueror root, is used as a lucky sex charm or mojo. Veterinary students, particularly the girls, appear to favour the *os penis* of the dog as a charm and this can be seen in use in Bristol today. From a historic point of view the mandrake root and the rhinoceros horn have provided the most potent sexual symbols and have thereby gained reputations as powerful aphrodisiacs.

At one time an inexplicable failure to procreate was often ascribed to some malign influence or to the direct intervention of an enemy or competitor. Charms were therefore often used purely as a prophy-lactic measure to counteract these influences. Many of the superficially light-hearted traditions associated with the modern wedding ceremony can be traced back to their origins in pre-Christian rituals. Since knots can protect against sorcerers, a net or fine veil with its concentration of knots can protect the wearer. The bridal veil therefore acts to keep the bride free from malign influences. On the other hand it was believed in some areas that a knot tied by a sorcerer could cause impotence in the bridegroom so, at the time of the wedding ceremony, care was taken to undo every knot in the clothing of both partners just in case such a charm had been cast. Both horsehoes and rice are considered lucky, and the latter is a clear fertility symbol. The bouquet thrown to the fortunate bridesmaid who catches it will ensure a prompt and fruitful marriage. Even the traditional timing of weddings to coincide with Easter or Whitsuntide is to gain advantage of Nature's time of maximum fertility at the vernal equinox. It is a strange paradox of modern society that we cling to the vestiges of early fertility rites in a ceremony which, in most cases nowadays, has non-fertility as its immediate aim.

Certain parts of the body have always been credited with mystic virtues or properties because of their recognized involvement with birth and procreation. The Greeks believed that the liver was the source of all generative power, and a recognition of the aphrodisiac potency of animal liver has persisted to the present day in parts of Europe. The idea that eating the generative organs of the obviously more potent beasts such as the stag, bull, or stallion as a means of transferring that animal's characteristics to the consumer has been practised in most human societies since prehistory. The early cave-dwelling hunters who drew representations of the hunt on the walls of their caves clearly believed that the hunters could gain the strength and ferocity of the animals they hunted by eating parts of their prey in special ritual ceremonies. In fact one of the motives behind cannibalism was not merely to take advantage of a readily available source of meat, but to gain the physical and mental qualities of the victim. The consumption of the brain was considered singularly important in this last respect.

Throughout recorded history it is possible to find evidence that any part of the anatomy of either sex of virtually any animal species which ancient thought could connect with conception or parturition was liable to become a sovereign remedy for disease, a reputed aphrodisiac, or an ingredient in a magic charm. Among domestic animals the cockerel has yielded a wealthy source of marvellous nostrums. A pregnant woman wishing to be assured of producing a son had only to eat the testes of the cock during her pregnancy. If one intended to take advantage of the undoubted aphrodisiac potency of the same organs, then it was necessary to wear the right testis, wrapped up in a ram's skin, as an amulet. The left testis, however, was liable to produce the opposite effect. A similar anaphrodisiac effect could be expected if the testes of an ordinary rooster (as opposed to a fighting cock) were placed beneath the bed together with some of the bird's blood. Clearly it was always wise, on your wedding night, to check carefully for untoward knots or suspicious residues underneath the wedding bed.

In the past it was presumably very easy to blame a temporary sexual inadequacy on an evil charm or curse of this type, but naturally the reverse was also true; a man wishing to ensure success in a sexual encounter would often enlist the aid of a witch or herbalist. The aid which would be provided, at a price, usually consisted of a charm of some considerable complexity so that any failure in the outcome could be put down to the simplest mistake in a vital part of the ritual. For example, to be certain of the efficacy of a pair of hyena's testicles, it was essential that the animal should be captured at the time the moon was passing through the sign Gemini. In Roman times, and even more recently, a menstruating woman was considered to be extremely unlucky. Her mere presence could produce the most dire consequences. According to Pliny it only required her touch to: 'render fruit trees sterile, wither the vine, empty hives of their bees, blacken linen in the wash, and cause pregnant mares to miscarry'. A girl who was menstruating for the first time could, by her glance, cause mirrors to tarnish, blunt steel, kill bees, corrode brass and rust iron, and it is remarkable that this catalogue of misfortune could ever have been believed. However, the same people who smile indulgently at these superstitions are probably not entirely averse to 'touching wood', 'throwing salt' or carefully avoiding walking under a ladder. With our present degree of enlightenment it is also easy to belittle the practice, common in Pliny's time, of burying a dog's genitalia in the foundations of a house to protect the household from enchantments. But many people are still prepared to nail a horseshoe over a threshold for precisely the same purpose, even if it is not openly admitted.

Turning from the corporeal to the celestial, the moon has always had a very important influence in magic and superstition. Many people still feel themselves affected to some extent by a full moon, and it holds romantic as well as aberrant associations. The word lunatic is derived from the Latin *luna*, the moon, in the sense of one who is controlled by the moon. Sexual offenders in particular often claim to be influenced by the cycles of the moon. The coincidence of the lunar cycle and the menstrual cycle of women would not have been lost on ancient magicians, and menstrual fluid always had great magical significance in early superstitious societies. In contrast, the male seminal fluid does not feature very strongly in the literature of the occult. The Romans, however, did believe that human semen provided a sovereign remedy for the sting of the scorpion, whereas the semen of the wild boar was relegated to the cure of earaches.

Aphrodisiac Charms

Several specific charms have lingered in folklore and it is worth considering some of them in detail. Some are strongly persistent in spite of all our modern educative processes. Perhaps at a time when the individual feels overawed or embarrassed by medical science, there is some excuse for the urge to try more ancient remedies in an attempt to return to Nature and things natural.

Hippomanes

The hippomanes was particularly valuable and versatile in its uses, although the exact nature of this substance is not clear. Tibullus describes it as the fluid which drips from the groin of a mare in heat, whereas Aristotle records various different charms all under the name of hippomanes. Virgil and Ovid describe it as a membrane or growth which is sometimes apparent on the forehead of a newly born colt, but which is very quickly bitten off and consumed by the mare. The *Shorter Oxford English Dictionary* is also unclear, giving the two alternative definitions: 'a small black fleshy substance said to occur on the forehead of a new-born foal; a mucous humour that runs from mares a-horsing. Both reputed aphrodisiacs.' The first definition suggests that it could be one of the possible origins of the unicorn myth. The horn of the unicorn, as that of the rhinoceros, has a long-standing reputation as an aphrodisac (see Chapter 5). Pliny's description of the efficacy of the hippomanes is, as might be expected, rather startling: 'If a person can remove this before the mare he has ready at hand a powerful philtre to engender passion in the frigid or indifferent. The mere odour of it will make an animal frantic.'

The practical use of hippomanes has long since disappeared and there is no zoological evidence for the existence of such a substance. Vaginal secretions, however, do contain several active substances, some of which might act as pheromones. This topic is discussed at length in Chapter 9.

Cockle-bread

This charm has been well described by the English diarist John Aubrey (1627-1697), who referred to it as 'a relique of Naturall Magick, an unlawful Philtrum'. It was employed by girls seeking to attract a young man and consisted of a small piece of dough which the girl would knead and then press against her vulva. The dough, moulded to this shape, was then baked in the normal way and the loaf presented to the man she sought to attract. If he ate it, he would fall beneath her spell and be powerless to resist. Similar types of charms have been used throughout Europe and indeed may still be used in primitive country regions.

Dragon's Blood

The efficacy of this substance in love potions and aphrodisiacs has been recognized for centuries. The 'blood of the mystic eye' mentioned in Egyptian magic texts is thought to refer to dragon's blood. To the Greeks it was cinnabaris (possibly because of its similarity in colour to the red sulphide of mercury, cinnabar, which was well known at the time and may even have been used as a cheap substitute for the real thing). Pliny, at his most credulous, states that it is the blood of a dragon or serpent which has been crushed to death by the weight of a dying elephant falling upon it. However, he adds, in all honesty, that it only resembles such blood, being in fact a gum.

Dragon's blood *is* a gum and can be obtained from either *Pterocarpus indicus* or *Calamus draco*, a species of palm (see Figure 3.1). It has a bright red colour and is used as a pigment in varnishes and enamels in order to obtain a crimson shade. It is for this reason that it has been available from chemists' shops and suppliers of artist's materials until quite recently. Its use in the ritual manufacture of a love charm has survived into this century. The most popular belief is that if a girl takes a small quantity of the gum, wraps it in paper, and throws it on a fire on Hallowe'en night saying: 'May he no pleasure or profit see till he comes back again to me', then he will, in due course, be obliged to return to her. A mixture of dragon's blood, quick silver (mercury), saltpetre and sulphur thrown on a fire while reciting a similar incantation is also claimed to be effective; it would certainly provide a spectacular firework if nothing else.

John the Conqueror Root

This is a popular twentieth-century charm found particularly in the southern United States. It consists of the dried root of St John's Wort (*Hypericum elodes*), which has a long and distinguished history as a magical plant. The root has an obviously phallic appearance and is placed in a small bag and worn around the neck. Unlike ginseng, which

Figure 3.1: The Dragon Tree, the Source of the Red Resin Known as Dragon's Blood. The drawing is based on an original illustration in Gerard's *Herball* of 1597

can present an equally phallic appearance but needs to be consumed in order to have any effect, John the Conqueror works magically by the Law of Similarity described earlier.

Magical Love Charms

Love charms are not strictly aphrodisiacs in that they are used to engender love rather than just venereal desire. It is certainly conceivable that certain substances are capable of increasing the sexual appetite or performance, but it is hardly likely that any recipe will be able to induce the complex emotions of love in an otherwise indifferent subject. For this reason, love charms tend to be magical remedies rather than active philtres. Medieval witches were herbalists as well as sorcerers and so would have been able to provide their clients with either love charms or aphrodisiac philtres to aid them in gaining the object of their desire.

Perhaps the most famous example of a love charm in fiction is that used by Oberon to bewitch Titania in Shakespeare's *A Midsummer Night's Dream*:

'Yet mark'd I where the bolt of Cupid fell:
It fell upon a little western flower, —
Before milk-white, now purple with love's wound, —
And maidens call it love-in-idleness.
Fetch me that flower; the herb I show'd thee once;
The juice of it on sleeping eyelids laid
Will make man or woman madly dote
Upon the next live creature that it sees.

Puck immediately sets off on his quest, and the use of this charm provides the plot for the rest of the play. The nature of this remarkable herb is thought to be the wild pansy or heartsease.

Some charms are rather less endearing. One particularly macabre example from Ireland is the 'dead-strip'. The girl who wished to gain the love of a man should go to a graveyard at night, exhume a nine-day buried corpse, and take a piece of skin from it. This strip is then tied round the arm or leg of the intended while he is sleeping and is removed before he wakes. This unpleasant ritual, it is claimed, will ensure the constancy of his love.

Fetishism also played an important part in many charms. The general principle consisted of obtaining some part of the woman's intimate clothing, preferably soiled with perspiration or blood, or, even better, a lock of hair or nail parings. These would be burnt to ashes in a special ritual and made into a drink by the addition of wine or mead. Once the woman had been inveigled into drinking this potion she would be under the power of the enchanter. Different rituals would be employed to invoke malign influences and misfortune upon the victim. It was this persistent faith in the effectiveness of such harmful charms that was largely responsible for the persecution of the witch cult in Europe and North America. The widespread belief in these methods is indicated by the frequent references to them by the writers and dramatists of the sixteenth and seventeenth centuries. Shakespeare, in *Othello*, makes reference to:

'. . . conjuration and mighty magic . . .'
'That thou has practis'd on her with foul charms;
Abus'd her delicate youth with drugs or minerals
That weaken motion'

Several medieval charms include rituals which are immediately recognizable from Egyptian, Greek or Roman practices, and it is quite remarkable how little these activities have changed over many centuries. There was obviously a strong tendency to keep exactly to the ancient traditions among the practitioners of these occult arts.

A very simple charm consisted of inscribing the letters – H, L, D, P, N, A, G, U, upon the left hand of the lover: 'Carry them in the morning before sun-rising and touch whom thou wilt and she will follow thee.' For the woman, the letters H, L, N, P, M, Q, U, M, inscribed in the same way before sunrise, would have the same effect if she surreptitiously touched the neck of her intended.

Charm rings with magical properties have been employed in England for curing various diseases from late Saxon times almost to the present day. They were usually made of copper or precious metals such as silver or gold and, as 'cramp' rings, were singularly efficacious if they had been blessed by the reigning monarch. A love charm of the sixteenth century requires that:

'One take two cramp rings of gold or silver and place them both in a swallow's nest that buildeth in the summer. Let them lie there nine days then take them and deliver the one to thy love and keep the other thyself.'

Rings used in this way often had magic words engraved upon them. This apparently increased their potency and influence. The famous ring of King Solomon was reputed to have the mystic word 'schem-hamphorasch' engraved upon it. The more complicated charms were intentionally made difficult to carry out so that failure to gain the desired effect could be blamed on some error or omission in the ritual. For example:

> 'Take three hairs of his head and a thread spun on a Friday by a virgin, and make a candle therewith of virgin wax four square, and write with the blood of a cocksparrow the name of the woman, and light the candle, whereas it may not drop upon the earth and she shall love thee.'

Another charm, whose origins can be traced back to ancient Greek times, runs as follows:

> 'Make an image of her you love in virgin wax, sprinkle it with holy water, and write the name of the woman on the forehead of the image and thy name on her breast. Then take four new needles and prick one of them in the back of the image, and the others in the right and left sides. Then say the conjuration. Then make a fire in her name and write on the ashes of the coals her name, and a little mustard seed and a little salt upon the image, then lay up the coals again, and as they leapeth and swelleth so shall her heart be kindled.'

A similar charm from the thirteenth century runs:

> 'Take the hairs of the woman whose love you desirest, and keep them until the Friday following, and that day before sun rising. Then with thine own blood, write thine own name and her name in virgin wax on parchment, and burn the hair and letters to dust on a red-hot fire, and give it to her in meat and drink, and she shall be so much taken with thee that she shall take no rest.'

The use of names in these charms is interesting since a person's name has always possessed mystical significance. Magic does not discriminate between words and the things they describe, so it is believed that the link between a name and the person to whom it belongs is not merely an arbitrary association, but a real bond in so far as the name becomes an integral part of the person as much as his hair or nails. As a result, a sorcerer can use one's name to invoke evil just as readily as if he had some other part of the corporeal body with which to perform his rituals. This explains why primitive people are peculiarly sensitive about revealing their names to strangers. A Brahman child, for example, receives two names, one for common use, and the other a secret name which only his parents should know. Similar concerns are expressed towards portraits or photographs. If either of these were to fall into the wrong hands, then it is possible that the most injurious magic could be wrought upon the unfortunate subject.

Several love charms involve the use of apples. The apple has become synonymous with sexual temptation from the story of the Garden of Eden in the first book of the Old Testament. One charm uses the symbolism of the apple together with elements of demonology as follows: 'Write on an apple: Guel + Bsatirell + Gliaell +, and give it to her to eat.' Alternatively, 'Write on an apple: Raguell, Lucifer, Sathanus, and say, "I conjure thee apple by these three names written on thee, that whosoever shall eat thee may burn in my love".'

The use of magic charms and rituals for procuring love or for ensuring faithfulness still survives even in the most advanced industrialized societies, and some of these ancient rituals seem to have lingered in children's rhymes. There is nothing remarkable about a child today pulling the petals off a flower and chanting 'He loves me, he loves me not, he loves me, he loves me not . . .' as each petal falls. Perhaps in the past this little game had a more significant purpose in the process of courtship.

Belief in magic rituals and charms was by no means restricted to Western European and mediterranean cultures. There are few surviving anthropological commentaries from before the last century, but the early explorer and traveller Engelbert Kaempfer, a German physician who lived between 1651 and 1716, has left accounts of the use of magic potions in oriental medicine. Kaempfer's medical training clearly left him with a healthy scepticism towards the magical side of medicine, although he retained a great respect for some of the other aspects of oriental medicine. For example, he writes:[1]

By administering a certain drug an adulterer can so blind a husband that he may with impunity climb into the husband's bed even though he is present. This is reported by the most respected Manelslo and Linschoton [who were apparently more credulous]. During my stay in various realms of India, I did not have an opportunity to test the truth of this claim. I did learn however, that marvellous phenomena and fantasies can be created in the brain which ignorant men believe to be magic worked with the aid of a spirit.

Like d'Orta before him (see Chapter 7), Kaempfer observed the effects upon behaviour of various intoxicating drugs including nicotine, opium, cannabis, betel nut (arecoline), and datura. On the subject of the sexual magic spells of the Makassars (the inhabitants of the Celebes Islands), who believed that spells and incantations could produce male or female impotence, he states:

> It is wrong to assert that bindings, locked doors, words muttered by witches, and other acts can induce impotence. An exception may be the case of a man of a superstitious nature, who through the workings of his own imagination might easily impede or promote the flow of spirits requisite for the sex act. In the case of newly weds, who are often bound and restrained from their marital function by a sense of shame, experience and familiarity loosen and restore that function. Those who try to explain effects of this type as being generated by the determination and mental cunning against another person by a spell caster do not distinguish between permanent and transient states.

This represents a remarkably sound view for the period in which it was written, and is evidence of a scientific mind unswayed by the myths and superstitions of the time. For the entertainment of his readers Kaempfer provides the following account of some sexual magic exactly as he obtained it from a practising witch:

> 'The girl who is to bind her lover, or the wife who is to bind her husband, must wipe his organ with an undergarment or any cloth after the sex act, collecting on it as much semen as possible. She must fold the cloth carefully and bury it in the earth beneath the threshold of her home. And as long as the cloth remains buried there, his organ obeys the will of no one but the woman who casts the spell. Nor will the man be freed from this bond until the cloth is dug up from beneath the threshold.'

On the other hand, if a man wishes to bind his bed companion he must burn one of her undergarments or any cloth well stained with her menses. He must then knead the ashes with his own urine into a phallic image. If there are not sufficient ashes for making even a small image, he must knead them with some earth on which he has recently urinated. Next, he carefully moulds the image and lets it dry. It must be stored in a dry place and protected from all moisture. As long as he preserves this model, anyone who attempts the sex act with this man's woman will instantly be impotent. But when the spell caster wishes to enjoy his own sexual pleasure, he must first moisten the magic model, and as long as it is wet the spell will be broken and he will be potent, but the same is true for all other men whom the woman allows to have sexual intercourse with her.

This magic chastity belt is certainly far more humane than the crude mechanical devices used in medieval Europe, and the even less pleasant practice of infibulation which is still carried out in certain Islamic countries. At the present time, people are not above using a little bit of magic, for example when swinging a pendulum over the body of a pregnant woman in order to determine the sex of her unborn child. In fact, there is still plenty of opportunity to take advantage of the credulity and ignorance of people as far as sexual activities are concerned, as will be made evident in the next chapter.

4 HERBALISM AND QUACKERY

The juxtaposition of herbalists and quacks is in no way intended as a slight to the former group of practitioners. Herbalism is an ancient and respected science which has led to the discovery of many of the drugs used clinically today. Leaving aside the question of the effectiveness of many herbal remedies, there is no doubt that the genuine herbalists have always had complete faith in the value of their treatments, and their principal objective has been to help their patients. The quacks, in contrast, have unashamedly sought to exploit their clients by offering nostrums which they know to be ineffective, and selling them at inflated prices. Preparations which are claimed to restore virility or banish sexual weakness are obviously going to provide a lucrative business for these charlatans who trade upon the gullibility and ignorance of their clients. Their activities have already been mentioned earlier, in the discussion of the role of advertising in the sale of aphrodisiacs. In this chapter the secret remedies of the quacks will be investigated in more detail.

There is a grey area between herbalism and quackery into which many of the reputed aphrodisiac preparations fall. The simplest means of differentiating between the two is to consider the intent of the seller as being the important criterion. Once a herbal remedy has been shown conclusively to be of little or no worth, then it becomes quackery to continue to sell it with that knowledge. The reason why many of the traditional nostrums contained 'secret' ingredients was not because of fear of competition, but because the ingredients were known to be both useless and a lot cheaper than the cost of the remedy would suggest.

It is fascinating to observe how the classic herbal texts have provided the basis for the use of drugs in medicine and how the materiae medicae gradually evolved into the modern pharmacopoeiae. The Merck Index, which is currently one of the best catalogues of chemical substances and is universally respected, had, as recently as 1905, a section devoted to an odd collection of substances described as aphrodisiacs. Aphrodisiacs no longer feature as such in drug lists, and the removal of the less savoury items from the official pharmacopoeiae was virtually complete by the 1930s. However, it does serve to indicate how recent has been the advance in our scientific knowledge concerning drug action and how close we still are to the Dark Ages.

Herbalism

The earliest herbal remedies appear in the *Pen T'sao Ching* (Great Herbal) of Emperor Shen Nung, which was written about 2700 BC, and the Ebers papyrus, which dates from 1550 BC. In Europe, the *De Materia Medica* of Dioscorides, written about AD 60 provided the principal basis for all the subsequent herbal literature for over 1500 years. The Saxon herbals, which provided the main body of English herbal lore, contained some empirical information, but were still largely derived from the works of Dioscorides. Even Nicholas Culpeper, in the most well known of seventeenth-century English herbals, acknowledges his reliance on the works of Pliny, Aristotle and Dioscorides.

The present practice of herbalism in Britain can be traced back to the principles set out by the Saxon writers. In fact, many of the names of plants are derived virtually intact from the original Anglo-Saxon: the suffix 'wort' is from *wyrt*, meaning herb or root, and gives rise to the term 'wortcunning', the study of herbs. The names we use today for many plants can therefore reveal their original uses in herbalism: for example, woundwort, heartsease, lungwort, liverwort and so on.

In the application of herbs to medicinal use, two important principles were observed, first, the Doctrine of Signatures, which stated that every plant that was of use to man had been marked by God in a way which revealed its intended use. This applied particularly to the organs of the body, so that if any plant, or part of a plant, resembled in shape, form or texture any part of the body (heart, lungs, liver, etc.), then that plant was indicated for the treatment of diseases affecting that organ. It was on this basis that many plants of phallic appearance, such as the carrot or asparagus, for example, gained reputations as aphrodisiacs. The mandrake root, which could resemble a whole man in terms of limbs, torso and head, was therefore believed to be a universal remedy and singularly powerful as a consequence of this similarity. This principle was not confined to the European herbalists; the Chinese had, and still have, an equal faith in the ginseng root for the same reasons.

The second important principle in the use of herbs was the influence of astrological signs both on the power of the herb and the course of the disease. Some of this starcraft was no doubt intended to compete with the religious quackery and faith in holy relics as practised by the established Church of Rome whose priests were in direct competition with the herbal physicians until after the Reformation. By that time, belief in the magical aspects of herbalism had considerably diminished, although Culpeper himself was not above the use of astrology to add some element of mystery to his herbal. Culpeper was undoubtedly intelligent as well as being an entrepreneur, and it is difficult to believe that he really had much faith in the more absurd and contradictory aspects of his astrological theories.

Culpeper's herbal has recently undergone a revival and several reprintings; its full title is *The English Physician Or an Astrological Discourse of the Vulgar Herbs of this Nation Being a Compleat Method of Physick whereby a man may preserve his Body in Health, or cure himself, being sick, for three pence charge, with such things only as grow in England, they being most fit for English Bodies*. In other words, it was the home medical encyclopaedia of the seventeenth century.

Some element of magic and mystery also helped to overawe the patient and perhaps even aid the cure. The power of faith and the level of expectation for the healing process should not be underestimated. Some important drug discoveries such as aspirin and digitalin have been derived from ancient herbal remedies, although it must be admitted that none of the enormous number of reputed aphrodisiacs has been considered worth subjecting to rigorous scientific scrutiny.

Witchcraft and the Church

The Saxons were knowledgeable herbalists but also had great faith in the use of magic to make love charms. The existing Saxon manuscripts of wortcunning, leechdoms and starcraft make this abundantly clear. An account of their practices published in the last century[1] reads as follows:

> The wizard, witch, sorcerer, druggist, doctor, or medicine man was equally ready at seeming affection. He played the part of a sort of ochreous Cupid. Instead of smiles and bright eyes, his dealings were with some nasty stuff put into beer, or spread slyly on bread. I have read somewhere of some agency known to Threofrastus not less potent than Spanish Flies, but if Saxon poisoners used them, they held their tongues about it.

In Cockayne's warning against witchcraft it is expressly charged that some women: 'work for their wooer's drinks or some mischievous stuff, that they may have them for wives'. Perhaps this Victorian author is doing these practitioners an injustice, although it is quite probable that the same individual who prepared love potions would have undertaken the manufacture of a more lethal brew if it were required.

In the Schrift Book of Ecgbert, Archbishop of York, one of the contemporary practices is strongly censured, and was subsequently considered to be so filthy that it was left in the obscurity of the original old English in later translations.[2]

Several herbs were employed to counteract the effects of love charms: betony, for example, was recommended by no less an authority than St Hildegard. Several Saxon|herbals, notably the Leech Book of Bald and the Lacnunga, have survived at least in part to the present day. Reading them makes it apparent that herbal medicine was already quite advanced by Saxon times, and the properties of *Belladonna* (deadly nightshade) and *Hyoscyamus* (henbane) were well documented. St John's wort, valerian and vervain were all used in love charms, but since the plant itself was not necessarily consumed in order to effect the charm, it would appear that it was the magical properties of the herb that were being employed rather than any endogenous active principle.

The fact that both herbalists with good intentions and witches with less certain motives used potent herbs in their specifics led to an increasing suspicion of self-professed herbalists, particularly during the periods of the great witch-hunts in the fifteenth and sixteenth centuries. At the time the church was preaching that all bodily ills were acts of God and were an indication of His displeasure at some sinful thought or deed. The only possible cure was therefore through the power of prayer and the intercession of the priest. Masses could be sung and prayers offered (for a price), and for the wealthy it was often convenient to pay for masses and prayers in advance as a prophylactic against future sinfulness and its retribution. It was also possible to pay for monks to undergo the penitences on the sinner's behalf. These facilities, needless to say, were not available for the poor, and despite all the attempts of the established church to prevent it, the craft of the herbalist and the village wisewoman survived and flourished.

Some of these early herbal remedies have since been found to possess a sound medical basis, the best example perhaps being the use of foxglove (*Digitalis*) as a cure for dropsy, the swelling of the tissues resulting from cardiac failure. In fact, it is likely that the peasants received more useful treatment from their herbalists than did the wealthy classes from the opportunist priests and charlatans who swarmed about them.

In England, the power of the Church of Rome and the sacerdotal quacks was largely removed by the Reformation. Many of the so-called holy relics which had been supposed to achieve the most miraculous cures were found to be fraudulent, and the way became clear for a more reasoned and scientific approach to therapeutic medicine to develop. Organized medicine at the time of the Reformation was totally dependent upon physicians and surgeons whose skills were very limited, and on the apothecaries whose knowledge was still mostly culled from the Saxon leechdoms.

Not surprisingly, the essentially superstitious nature of the people was not going to be changed overnight, and in the face of the general incompetence of the formal physicians, they turned from the discredited holy relics to the host of quacks and charlatans who proliferated from then on. There was no knowledge of antisepsis or anaesthesia, and the payment of a high fee to a quack for a painless, if ineffective, remedy must have seemed preferable to an extremely painful and possibly fatal experience at the hands of a self-taught surgeon. Against this, the herbalists, who could at least derive a soporific and sedative mixture from henbane and belladonna, were bound to be in demand. Many of the village herbalists were old women who also acted as midwives and were instrumental in keeping alive the old superstitions concerning love charms, sex and childbirth. It was also at this time that the pressure from the new religious reformers caused the first penal laws against witchcraft to be placed on the statute book, and many of these old midwives ran the considerable risk of being accused of witchcraft as a consequence of their practices.

However, as communication between European countries improved and the wealthy elite had more time to spend on romantic exploits, the market for love philtres and aphrodisiacs increased. The Doctrine of Signatures was still invoked in the continued belief that plants with any phallic features were certain to be aphrodisiac in their effects. Asparagus, carrots and parsnips could be included for this reason, asparagus particularly since it had the advantage of being relatively scarce. The initial scarcity and value of tomatoes and potatoes immediately following their introduction from the New World into Europe was a significant factor contributing to the early belief in their aphrodisiac properties. Only the wealthy could afford such luxuries and it is not surprising that the poorer classes believed that the all-too-evident promiscuity of this fortunate minority was due to their consumption of such rare and exotic delicacies rather than to their generally higher standard of living and increased opportunity for sexual pleasure.

'Exotic' and 'rare' are still qualities that are expected from any potential aphrodisiac. One has only to look at the illicit trade in rhinoceros horn from Africa into Asia today to appreciate this point. The sixteenth and seventeenth centuries were times of great exploration, and the discovery of new lands meant that various exciting new tropical fruits and spices became available for the first time for culinary and other purposes. They were soon included in the armamentarium of the manufacturer of aphrodisiacs. The development of the hothouse and improved means for the rapid transportation of delicate vegetables means that there are very few truly exotic or rare fruits and spices today which cannot be obtained fairly readily. Nevertheless, there still lingers the basic willingness to believe in the efficacy of the rare and expensive, as is well shown by the recent 'discovery' of Pega Palo (see page 92). That this type of myth can occur today in an educated and supposedly materialistic society should perhaps make us more tolerant towards the apparent naivety and gullibility of the first Elizabethans.

Vegetable Aphrodisiacs

The Tomato

The tomato or love apple was a scarce and expensive delicacy for some time after its discovery in America and introduction into England. The name 'love apple' illustrates how easily names can become distorted through mistranslation. The original Latin name for the tomato was *mala oethopica* or 'apple of the moors'. In Italian this became *pomi dei mori*, and then was subsequently corrupted on translation into French as *pomme d'amour*. In English this was rendered as love apple.

The colour of the tomato also enhanced its reputation as an aphrodisiac: such a large, bright-red fruit was unusual at that time and red was a colour generally associated with poisonous fruit. In addition, red has always been the colour linked with passion and excitement. The increasing popularity of the tomato as a vegetable soon meant that it was being grown on the continent of Europe and in the Channel Islands for export to England. By the nineteenth century, when it had become plentiful and cheap, it no longer attracted any attention as an aphrodisiac. Medicinally it was considered to be good for promoting bile and as a protection against cancer, although it was also recognized that the consumption of uncooked green tomatoes could produce the 'colick'.

The Potato

The potato (*Solanum tuberosum*) was first introduced from Peru in 1586 by Thomas Heriot and only later was it introduced into Ireland from Virginia by Sir Walter Raleigh. Early references to the potato are almost certainly describing the sweet potato or *batatas*. This tuber, like the root of the sea holly (see below), had a reputation for restoring the waning vigour of the ageing reprobate. This is why Falstaff is made to exclaim:

'Let the sky rain potatoes, hail kissing comfits and snow eringoes.'
Merry Wives of Windsor Act 5 Scene 5

For a long time the potato was kept as a delicacy among the wealthy, and it was only in the nineteenth century that it became the cheap and staple diet of the working classes, to the extent that in Ireland the failure of the crop resulted in widespread starvation and the enormous exodus of Irish people to America. The potential of the potato as a source of alcohol was not missed by early experimenters: potatoes were boiled with dilute sulphuric acid (which converted their starch to glucose), and then fermented with yeast to yield an alcoholic spirit sold as 'British brandy'. It probably tasted as awful as its name would suggest.

Sea Holly

The sea holly (*Eryngium maritimum*) is a rather scarce plant found growing wild around the coast of England. At one time it had a considerable reputation as an aphrodisiac, as recorded in Dryden's translation of Juvenal's *Satire VI*: 'who lewdly dancing at a midnight ball for hot eryngoes and fat oysters call'.

The herbalist Robert Burton even established a factory at Colchester for producing the candied roots of the plant. These were very popular in the sixteenth and seventeenth century and were sold as candies known as kissing comfits. Gerard, in his herbal, states:

The roots if eaten are good for those that be liver sick; and they ease cramps, convulsions, and the falling sickness. If condited or preserved with sugar, they are exceeding good to be given to old and aged people that are consumed and withered with age, and which want natural moisture.

During the Elizabethan era it was believed that the roots, when taken in moderation, would restore masculine vigour and the vital spirit. Lord Bacon, recommending a concoction of malmsey, wine and eggs, concluded:

> 'You shall doe well to put in some slices of Eryngium roots, and a little Ambergrice; for by this means, besides the immediate facultie of nourishment, such drinke will strengthen the back.'

In the United States another species of the eryngo, the buttonsnake root (*E. aquaticum*), was widely used for its medicinal properties. Its roots were claimed to excite the venereal desires and strengthen the procreative organs. It featured in a number of quack remedies. It is apparently still employed in some homoeopathic remedies for diseases of the bladder.

The Quince

A jam or marmelo made from this highly fragrant fruit is thought to be one of the possible origins of the word marmalade. The quince seems to have lost popularity over the last 200 years although, as the author can testify, the fruit produces an excellent white wine with a distinctive scented taste. At one time the quince was a great delicacy, although it will grow quite readily in the southern areas of Britain. The fruit was usually candied or made into a jam. It was an essential ingredient of wedding feasts and, perhaps as a result of this tradition, the quince developed a reputation as an exciter of sexual appetite. Possibly it was with this in mind that the Lady Lisle sent her daughter Catherine some quince marmalade to give to King Henry VIII.

The Globe Artichoke

This is still a popular, if slightly expensive, vegetable in this country. The florets of the flower head are boiled and the soft fleshy part at the base of each floret is eaten with the appropriate seasoning. According to Gerard:

> This middle pulp when boiled with the broth of fat flesh, and with pepper added, makes a dainty dish being pleasant to the taste, and accounted good to procure bodily desire. (It stayeth the involuntary course of the natural seed.)

This last statement is presumably a reference to the possible prevention of premature ejaculation. It is also significant that at the time when

Gerard was writing, the globe artichoke was still imported at some expense from Italy and therefore possessed a rarity value in addition to any intrinsic potency. Sunflower buds cooked in the same way as the artichoke were at one time claimed to be even more effective in the provocation of lust.

Arum Lily (Arum maculatum)

This is a common plant in Britain and has a very large number of local names: Lords and Ladies, Cuckoo Pint, Wake Robin, Parson-in-the-Pulpit, Rampe, Starchwort, Arrowroot, Gethsemane, Bloody Fingers, Snake's Meat, Adam and Eve, Calfsfoot, Aaron, and Priest's Pintle represent merely a selection. Within these names it is possible to recognize thinly disguised references to the sexual organs. The red berries of the arum lily, which appear in the early autumn, are poisonous and extremely unpalatable, so it is not clear how the plant was supposed to be used. Possibly the Doctrine of Signatures was invoked to claim the effectiveness of the rather phallic male parts of the plant in stimulating the functions of the male organ. In the distant past a decoction of the leaves was used to loosen phlegm, and Dioscorides used the leaves mixed with cow dung as an external poultice for gout. The tuberous root was widely employed to prepare a powder from which starch could be made, and at times the tubers have even been eaten as food.

Wormwood

Wormwood is a generic term referring to several plants from the genus *Artemisia*, most notably *A. absinthium* which was once responsible for the distinctive flavour of absinthe. Since the natural oil of wormwood originally used in making absinthe is also a long-acting neurotoxic poison, wormwood was also responsible for a considerable number of deaths among regular drinkers. In France, where it was particularly popular, wormwood was banned and alternative bitter essences were found with which to flavour their traditional *pastis*.

Wormwood has a long history, being mentioned by both Pliny and Dioscorides. Pliny has been quoted as saying that wormwood was a cure for syphilis. As far as we know, syphilis was only introduced into Europe at the end of the sixteenth century, and earlier references to it are probably describing leprosy. However, Pliny also states that wormwood is a great sexual restorative: 'Haec etiam venerem pulvino subdita tantum incitat.'

Wormwood probably derives its name from its proven efficacy at expelling worms, particularly roundworms, and it has appeared in several pharmacopoeiae as an anthelmintic. This property is due to the active compound, santonin (1,2,3,4,4a,7-hexahydro-1-hydroxy-α,4a,8-trimethyl-7-oxo-2-naphthalene acetic acid *l*-lactone), which is still obtained by extraction from the dried unopened flower heads of *A. maritima*. Santonin continues to be used in veterinary medicine, but has been superseded in clinical medicine by less toxic vermifuges. The symptoms of santonin poisoning include alterations to the visual perception of colour (xanthopsia), headache, vertigo, nausea, vomiting, profuse sweating, and diarrhoea. Large doses can even produce epileptiform seizures, respiratory failure, and death. The drug has the unusual property of turning acidic urine bright orange in colour and alkaline urine purple. This must have been a rather alarming symptom to patients given santonin as a vermifuge.

The major aromatic component of the volatile oils derived from wormwood is thujone, a compound which is closely related to camphor. The properties of thujone have still not been thoroughly investigated, but there are suggestions that it may have some similar actions to cannabis.

Wormwood has been used for a variety of purposes by herbalists for over 2000 years. *Artemisia vulgaris* or mugwort was supposed to protect and strengthen travellers. According to Pliny these benefits could be achieved by merely tying the dried leaves about the waist or even holding it in the hand. It was also known as St John's belt in the belief that John the Baptist wore it when he was in the wilderness. In this context it would also protect the wearer against witchcraft, and houses against lightning. Wormwood also had more practical uses as a means of deterring lice and fleas. William Tusser, writing in his *Five Hundred Pointes of Good Husbandrie* in 1573, states:

'While wormwood hath seed get a handful or twain,
To save against March, to make flea to refraine,
Where the chamber is swept and the wormwood is strowne,
No flea for his life dare abide to be knowne.'

The use of wormwood as an aphrodisiac is only hinted at by the various writers on the subject. The alternative names for wormwood, such as Boy's Love, Lad's Love and Maiden's Ruin do suggest, however, that it had gained a certain reputation.

The French liqueur absinthe contains, in addition to wormwood, oil of marjoram, aniseed, dill and other aromatic oils. As a drink it was supposed to be aphrodisiac and was very popular in the artistic community in Paris at the turn of the century. There was more evidence, unfortunately, for its reputation as an inducer of epileptic fits and eventual madness. Its undoubtedly toxic properties can be ascribed largely to the cumulative effects of santonin and thujone, but it should be remembered that absinthe also contains about 40 per cent alcohol and is therefore a fairly potent drink by any standards. Alcohol is often the major ingredient in preparations that are claimed to have aphrodisiac properties, and absinthe is no exception.

Apart from mugwort, a very large number of flowers and herbs were reputed to be aphrodisiacs, and some of their alternative names provide indirect evidence of this usage. The ragwort, a fairly common and widespread weed (*Senecio jacobaea*), is a corruption of ragewort, meaning a stimulant of the sexual organs. It has also been known as cankerwort and St James Wort, and was considered especially efficacious for curing the staggers in horses. The sundew (*Drosera rotundifolia*) is one of the few insectivorous plants found growing in Britain; its alternative names were youthwort or lustwort, which should be self-explanatory. It was believed that female cattle would have their copulative instincts excited by eating even a small quantity of the plant. As a herbal simple it is used as a specific for whooping cough. The periwinkle (*Vinca minor*) was used as a purgative and an astringent to staunch bleeding, but it may have had other uses since, according to Culpeper: 'The leaves of the lesser periwinkle, if eaten by man and wife together, will cause love between them.' It is not clear whether or not Culpeper was referring to love in the spiritual or physical sense. The list of possible herbal aphrodisiacs could be continued at some length, but in the absence of any concrete evidence of a positive effect, they are probably best left to history.

Ginseng

'The branches which grow from my stalks are three in number,
and the leaves are five by five.
The back part of the leaves is turned to the sky,
and the upper side downwards.
Whoever would find me must look for the Kia tree.'

Korean song

Ginseng has been used for centuries in China, Tibet, Korea, Indo-China

Figure 4.1: Examples of (a) *Panax ginseng* and (b) *Mandragora officinarum*. The outward appearance of the roots of both plants shows a marked similarity, which may account for their similar reputations as cure-alls

and India, where it is reputed to possess the same properties as the mandrake root. Only recently has its use penetrated to the West, where it is now subject to extensive scientific and medical investigation. In Asia is can be bought in the form of teas, tinctures, ointments, pills

Figure 4.1 (b)

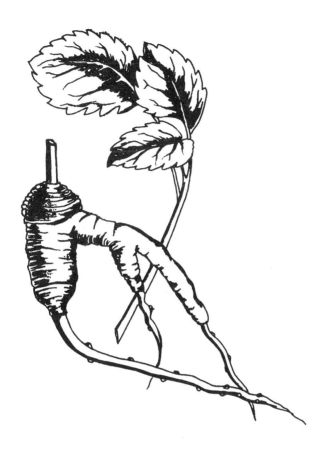

or powder, and is recommended for, among other things, tiredness, headaches, diseases of the liver, heart, kidneys, tuberculosis, and diabetes. It is also firmly believed to be an aphrodisiac with the power of rejuvenation.

The plant from which Asian ginseng is derived is *Panax ginseng*, a small tree or shrub of the Araliaceae family found in the shady forest regions of Manchuria, Korea and the USSR. Exports of ginseng roots have apparently provided a valuable source of income for the natives of these regions over a period of several centuries. The shape of the root can resemble the human form (see Figure 4.1a) and thus, by the Doctrine of Signatures, it should be universally beneficial to the body. In fact, the name of the tree can be loosely translated as *panax* = cure-all (as in panacea), and *ginseng* (or *schinseng*) = man-like. The root is remarkably similar in appearance to the mystical mandrake mentioned later in Chapter 5.

Mrs C.F. Leyel, in her *Modern Book of Secrets*, quotes Sir Edwin Arnold as saying:

'It [ginseng] will renovate and re-invigorate falling bodily powers beyond all other stimulants, stomachicks and energisers of vitality. The Korean people believe the said root to be an absolute panacea for all mortal ills, mental or physical; it is packed and transmitted with the most scrupulous care and pains, in small parcels of white silk, the mouth and nose of the recipient having to be covered when unfolding these sacred envelopes of embroidered silk or of crimson and goldfish skin. The habitat of this wonderful root (in form like a man) is in the glens and slopes of the Kang-ge mountains, and it can be found only by people of blameless life and purity of heart. When taken from the earth it is thought to utter a low musical cry [again like the mandrake]. It is to be cooked in a special silver kettle having a double interior, as an infusion or with rice wine. From sixty to ninety grains of the dried root are a proper dose; it fills the heart with hilarity, whilst its occasional use adds a decade of years to the ordinary span of human life.'

An American species of ginseng (*Panax quinquefolium*, the five fingers root) was discovered in 1718 near Montreal by a Jesuit missionary who happened to be familiar with the miraculous properties of the roots of the Asian variety of *Panax*. There followed an immediate 'goldrush' to the area and thousands of prospectors started digging for the roots. The price of ginseng rose dramatically in Quebec and, for a brief while, a lot of money was to be made. Eventually, the indiscriminate harvesting of young plants led to its virtual extinction in that region of Canada.

More recently, the plant has been cultivated on plantations in South Korea, Japan and the USSR, where there is considerable official interest in the exploitation of the properties of ginseng. Most of the ginseng available in shops in Britain originates from South Korea. At one time it could be found only in speciality health-food shops, but a recent advertising and marketing campaign has resulted in the appearance of ginseng tea in most large supermarkets and, although it is not cheap, the consumption of ginseng is growing very rapidly both here and in the United States.

The Properties of Ginseng

An ancient Chinese pharmacopoeia claims that ginseng will 'allay fever, expel effluvia, brighten the eyes, open the heart, invigorate the body, and prolong life'. The current advertising of ginseng, however, has to be more moderate in its language, and no direct claim for a rejuvenating action can be made. Nevertheless, an increasing number of people seem to be convinced that ginseng is doing them good. Whether or not they are correct in this view is a matter that will be considered in more detail later. The aphrodisiac properties of ginseng have been implied from its use throughout history as an invigorating and rejuvenating agent of established value. For the ageing sybarite who wishes to preserve and prolong his powers, ginseng undoubtedly holds considerable interest.

Faced with a plant reputed to possess a wide range of physiological properties, the recognized scientific approach is to make extracts from it in order to separate, identify and purify the active principles. For example, the cardiac glycoside digitalin is still extracted from foxglove plants grown on farms run by the pharmaceutical companies. Using this approach the ginseng root has been found to contain glycosides, saponins or panaxosides, fatty acids, amino acids, sugars, cellulose, volatile oils, and the trace elements copper, cobalt, manganese, germanium, and vanadium. The miraculous powers of the root could theoretically reside in any or all of these substances, and the next logical step would be to test each component in turn. The traditional Asian approach, on the other hand, is to regard the root as a whole rather than as a source of more potent material. Indeed, the Chinese and Indian textbooks state quite specifically that, once ginseng is broken down into its various constituents, the singular beneficial activity of the root is lost.

Research into Ginseng

Research into the precise actions of ginseng were originally hampered by the lack of a standardized extract of the root that could be used in controlled experiments. The occult herbalists might claim that the potency of an individual root would depend solely on the degree to which it resembled the human form. In contrast the modern pharmacologist would argue that the potency of a root depended entirely upon its content of active principles, and that this in turn would depend upon such factors as the growing conditions, the season of the year, the age of the plant, and the conditions under which the root was stored after harvesting.

Independent experiments carried out in laboratories in the USSR and Japan over the last two decades have tended to confirm that ginseng does indeed contain some factor(s) that can increase stamina and endurance. In a human study, the gerontologist Medvedev[3] found that tired radio operators made fewer mistakes in transmitting messages after a drink of ginseng, when compared with a control group who were given an indistinguishable drink lacking ginseng. The experiment was well designed in that the radio operators themselves did not know which drink they had been given. Although these and other experiments have shown that ginseng possesses a significant stimulant action, it is very interesting that the subjects taking part in these experiments did not appear to be aware of any physical side-effects of the ginseng. All the other well-known stimulants such as caffeine, cocaine and amphetamine produce very obvious adverse reactions such as insomnia, hyperactivity, throbbing headache, palpitations, dry mouth and so on. In addition to these drawbacks, the stimulants as a group also possess a high risk of physical dependence. This does not seem to be the case with ginseng.

Another beneficial feature of ginseng is its remarkable lack of toxicity; the dose required to produce toxic symptoms in man has been reported to be about 1000 times the normal effective dose. This would be equivalent to consuming between 1.0 and 1.5 kg of root and, since the current price for a 30 g packet of ginseng extract is about £15 or $18, the total value of this massive dose would be between £450 and £675! This represents a wider margin of safety even than caffeine, which is present in tea and coffee at a concentration of between 1 and 3 per cent by weight and can cause adverse reactions in sensitive individuals. This lack of toxicity implies that ginseng, or at least one of its constituents, is acting as a stimulant by some mechanism different from that of the other stimulants. Although the Americans are renowned for their high intake of coffee as a means of remaining alert, the North Vietnamese delegation at the peace talks held in Paris during the Vietnam war claimed that their own astonishing stamina and endurance were due to the regular consumption of ginseng.

Sex and Ginseng

If we accept that ginseng is a potentially useful and safe stimulant, then it is perfectly logical to assume that it will also be capable of stimulating the intellectual and physical abilities of the individual in a sexual direction. However, there have been no reports of any direct aphrodisiac effects of the root, in spite of its suggestive shape. It is sometimes worn as an amulet or charm, in rather the same manner as

John the Conqueror root, and can perhaps act as a useful conversation piece. One drawback of the drinking of ginseng tea for its aphrodisiac effects is the recently increased availability of the material. All sense of mystery and excitement tend to be lost when the object of so much oriental myth can be found in a department store such as Harrods, alongside tea, coffee and drinking chocolate. It was, in fact, the very popularity of chocolate in the eighteenth century which led to the loss of its aphrodisiac reputation. After all, once it had become cheaply and easily available to the common people, it could hardly even be considered fashionable, never mind aphrodisiac. Perhaps this is the fate that will eventually befall ginseng.

From the scientific point of view, there have been a number of animal experiments in which ginseng extracts have been found to produce changes in hormonal balance and sexual activity. In one study the mating behaviour of male rats was measured, following the treatment of the animals with a tincture of ginseng.[4] The rats were paired with receptive females and the following responses were observed: mounting latency; intromission latency; inter-intromission period (i.e. how often); ejaculatory latency; success rate (the proportion of mountings resulting in successful intromission). The remarkable sexual capability of the laboratory rat may be judged from the fact that all the parameters listed were measured in seconds rather than minutes. The pretreatment of the male rats with ginseng extract caused a shortening of the latency to ejaculation and, at the same time, increased the rate of ejaculation during the 45-minute observation period. The incidence of mounting did not differ between the control and ginseng-treated groups, and so it appeared that the effect of the ginseng was confined to the ejaculatory activity of the rats.

On the basis of these observations the authors concluded that ginseng in some way facilitated the copulatory behaviour of the male rat. In some earlier experiments it was the females who received the ginseng, and they subsequently became more responsive to the advances of the males when compared with an untreated group of animals. Other groups of workers have detected increased ovarian growth in frogs, and enhanced egg-laying in hens given ginseng. Although the last observation might be of interest to farmers with battery hens, it is difficult to appreciate the immediate relevance of these experiments to the human condition.

It is considered highly unscientific to extrapolate, without reservation, the results of experiments on animal behaviour to human behaviour. Sexual behaviour in animals is very much more instinctive than in humans, and drugs tend to have more clear-cut and less complex behavioural effects in animals. This limitation in the interpretation of animal behavioural data has not prevented some individuals from experimenting upon themselves with extremely toxic drugs which have been shown to increase sexual activity in rodents. One particular example was para-chlorophenylalanine (PCPA), which had a brief period of notoriety among the drug community of California in the late 1960s.

Current Use and Abuse of Ginseng

It was mentioned earlier that ginseng was unusually non-toxic, with few if any adverse actions. However, as its use has increased, there have been a growing number of occasions on which adverse reactions have occurred. In North America, ginseng is available as whole or powdered roots, in the form of tablets, tinctures, capsules, infusions, and even as sweets or in chewing gum. It is promoted as part of the back-to-nature 'naturopathic' way of life, and consequently it has an appeal to the size-able minority of people who wish to reject the synthetic products of the pharmaceutical industry. This view also tends to be associated with the false belief that 'natural' products are somehow less toxic than man-made drugs, and that they can safely be taken in large doses. Even common vegetables such as carrots and tomatoes have been shown to produce symptoms of poisoning when eaten to excess, so it is not surprising that some cases of ginseng toxicity have come to light.

A recent survey of 133 regular ginseng users[5] reported the various untoward effects of ginseng as described by the users themselves. On the positive side, a large majority (93 out of the 133) reported feelings of well-being and increased physical and mental efficiency. Less pleasant symptoms included morning diarrhoea (47 cases), insomnia (26 cases), and skin eruptions (33 cases). The most disturbing observation was that 22 of the 133 subjects experienced a significant rise in their blood pressure that was actually measured by the researchers. If a commercially produced drug was ever found to be responsible for this incidence of side-effects, it would almost certainly be withdrawn instantly. But herbal remedies are not at present subjected to the same stringent controls as pharmaceutical products. In the same survey only nine subjects reported an enhancement of their sexual performance while taking ginseng. Assuming that these people were taking ginseng with some degree of expectation of a beneficial effect, this is a very disappointing result which casts doubt upon the efficacy of ginseng as an aphrodisiac.

The major limitation in this study of ginseng users was the lack of information concerning the quantities of ginseng that the subjects were taking, although it appeared to make no difference whether it was taken in the form of tablets, capsules or tea, by mouth, or by other unspecified routes. Also, no precise details were given in the report of precisely how sexual performance had been enhanced, nor whether this effect was limited to one sex. It is possible that, in the subjects who experienced an increase in blood pressure, some facilitation of the erectile process might have occurred. Other drugs which are used to alter the blood pressure do occasionally have this side-effect (see page 239).

In summary, it would seem that there is a low risk associated with the prolonged or excessive use of ginseng, but the relatively low incidence of aphrodisiac effects compared with the diarrhoea, insomnia, and general irritability also produced makes ginseng a poor choice as a means of improving sexual performance *per se*. Taken in moderation it probably does have mild stimulant effects and, for this reason, can be a useful, albeit expensive alternative to tea and coffee. The more extravagant claims made on its behalf almost certainly depend more upon the faith of the user than upon any intrinsic potency.

The Legacy of the Herbalists

Herbalism is undergoing something of a revival at present, possibly as a reaction against the myriad new drugs appearing on the market and the readiness of general practitioners to prescribe them. The fact that this revival also coincides with a rekindled interest in so-called natural foods suggests that society, or at least those sections of it able to afford the luxury of 'natural' products, are beginning to feel manipulated by the manufacturers and advertisers of high-technology convenience foods and are reacting against them. There is no doubt that the herbal lore, accumulated over centuries by different societies, was the springboard from which the scientific development of modern drugs took off. It is only necessary to read the textbooks of pharmacology (the study of how drugs act) published in the 1920s to appreciate how recent and how rapid has been the growth of knowledge of the mechanism of action of drugs on the body. From the appreciation of the fact that certain plants contain potent substances which can be used to cure disease, it is only a short step, given the growth in analytical chemistry techniques at the end of the last century, to the rational development of novel compounds with specific actions.

Modern homoeopathic medicine, although only put on a scientific footing by Hahnemann in the last century, has its origins in the medical writings of Hippocrates, Galen and Paracelsus. At a time when formal medical practice consisted largely of bleedings and purgings, homoeopathy provided a radical and occasionally successful alternative. The principles on which homoeopathy is based are disturbingly close to the spurious Laws of Similarity and Contact, which determine the practice of magic. However, there is no doubt that in some cases it does achieve successful cures where current medical procedures have failed.

Although we would now recognize that many herbal remedies do contain active principles, the rationale behind herbal remedies is quite different from pharmaceutical practice. Modern pharmacologists believe that it is essential to extract, purify and chemically identify the active component from a plant extract so that the potency of a preparation can be standardized and the dose of active drug known exactly. The herbalists, on the other hand, insist that the entire plant is essential for a natural and balanced effect; no single component will therefore be as effective as the whole. This difference of opinion has contributed to the controversy over the use of ginseng in particular. The early herbalists, using the principles of magic which we now recognize as having no sound basis, nevertheless discovered a range of useful drugs, many of which are still in use today. Unfortunately, none of the herbal aphrodisiacs has so far been considered worth pursuing on a scientific basis, one of the problems being that so many plants have been reputed to act as sex stimulants at one time or another.

Quackery

The increasing prosperity of Britain in the eighteenth century, mainly as a result of the industrial revolution, contributed to the religious fervour of the previous centuries being replaced by a more materialistic attitude to a life in which there was increasing scope for gaining some of the wealth that had previously been limited to the inherited fortunes of the landed gentry. This was the dawn of the age of the self-made man, which reached its peak in Victorian England. The climate was ideal for the proliferation of charlatans and quacks, many of whom made fortunes out of a gullible and ignorant public. Some of these plausible characters even received royal patronage: naivety and ignorance were not the prerogatives of the poor. Since, by definition, the quack methods for achieving increased sexual potency are inactive, there is little point in seriously assessing their effectiveness. What is far more interesting is to observe how these remedies and devices have persisted to the present day, and to examine the extent to which the search for sexual satisfaction has influenced social practices over the same period.

Medicinal Waters

> There is a Place adown a gloomy Vale
> Where a burden'd Nature lays her nasty Tail
> Ten thousand Pilgrims thither do resort
> For ease, Disease, for Lechery and Sport.
> (Earl of Rochester: *Bath Intrigues*, 1725)

One of the most enduring features of the golden age of quackery can be appreciated from the large number of spas that were discovered or rediscovered and whose names are still with us. There are undoubtedly many natural mineral springs which have beneficial medicinal properties, but there were also many whose properties were exaggerated in order to attract a wider clientele and, in some cases, it was the sexually restorative nature of the water that was emphasized.

Miraculous claims were made in order to attract the lucrative trade from rich patrons. Certain spas became fashionable for particular disorders, the spa at Bath, the most fashionable of all, being particularly recommended for its stimulating properties and power to revitalize jaded passions. Consequently, it very soon gained a reputation as the principal source for unbridled lechery among the aristocracy. Debauchery became commonplace, and the large itinerant population of prostitutes ensured that the local quacks enjoyed good business in selling cures for the French pox and other venereal diseases.

It was not so much that the waters of Bath actually contained any sexually stimulating elements, but rather than the social amenities provided a fine opportunity for like-minded people to congregate and indulge their pleasures to the full. Modern guidebooks to Bath spa make much of the health-giving qualities of the water (which can still be drunk in the famous pump room), but neglect to mention the more salacious aspects of the spa's past.

It was at about this time, when the spas were at their peak of popularity, that an Edinburgh physician, James Graham, introduced his 'Celestial Bed' to the London public at a 'Temple of Health' in Pall Mall. This remarkable piece of furniture measured twelve feet by nine, was decorated by extravagantly carved cupids, and rested on glass pillars which served to insulate the occupants so that their electrical fire could stimulate the generative process. Not only was the mattress stuffed with stallions' hair as an additional inducement to passion, but the entire bed could be tilted in order to facilitate copulation.

The object of the bed, apart from generating a healthy income for Dr Graham, was to ensure that the love-making of the occupants would be successful and result in conception. Barren couples could rent the bed for a not inconsiderable fee which included the hire of a small orchestra who provided appropriate music to accompany their physical activities. Graham's celestial bed was reputed to have cost him £10 000 and must surely hold the distinction of being the largest and most expensive aphrodisiac device ever invented. Unfortunately for Dr Graham, it was not a profitable venture and he eventually died in poverty in 1794.

The American Pitch Doctors

On the other side of the Atlantic the travelling medicine shows were a feature in the newly settled areas of the Mid and Far West. The principal object of the shows was to get hold of the money of the local people by persuading them to buy miraculous medicines at grossly inflated prices. The advertising techniques employed to further this end have already been described (see Chapter 1), but the other great art practised by these quack doctors, who very much relied on their wits to survive, was that of the sales pitch. Not all the members of their audience were fools, and in a society in which the carrying of guns and other weapons was commonplace, an argument with a dissatisfied customer could easily have fatal consequences. As a result the quack doctors also tended to be adept at keeping on the move and ensuring that their old clients did not catch up with them.

It is still possible to hear the occasional good sales pitch at a street market or travelling fair, and it is almost worth buying the goods in order to enjoy the salesman's banter. But the modern pitch is only a pale shadow of the methods employed by the unashamed quacks of the last century. Holbrook, the author of the book *The Golden Age of Quackery*, was fortunate enough to meet Mrs Violet Blossom long after she had retired from the medicine show business. He was given a demonstration of the 'Vital Sparks' lecture with which she persuaded men of all ages to part with a dollar for buckshot candy coated with aloes and worth less than five cents. Mrs Blossom was of

Chinese origin and called herself Lotus Blossom for the purpose of the lecture.

Lotus Blossom began by explaining that, in China, it was the custom for the secrets of medicine to be handed down through the female side of the families of the great mandarins. Thus, she alone possessed the full knowledge of the story she was about to relate. During the Whang Po Dynasty there occurred a sudden and inexplicable fall in the birth rate, to the extent that the Emperor offered a huge reward for a remedy which would restore the vitality of several hundred million Chinese men. Apparently all efforts failed until a certain He Tuck Chaw provided the solution to the problem. While exploring a remote volcanic region of Outer Mongolia he chanced to notice a vast population of small turtle-like animals. A very small proportion of these animals were remarkable for their beautiful golden stripes. He Tuck Chaw was intrigued as to the reason for the relative scarcity of these striped turtles and he settled down to investigate them. It turned out that they were the males of the species and that the ratio of females to males was about 1200 to 1. (There was a pause here while the significance of this ratio sank in.)

He Tuck Chaw therefore set about discovering the reason for the incredible vitality and potency of these precious few males. He found that they were distinguished by a small pouch at the base of the brain to which he gave the name Quali Quah pouch. He swiftly removed these pouches and made a powdered extract with which he returned to the Emperor's court. On being told of the discovery the Emperor provided He Tuck Chaw with some suitably decrepit old men on whom to experiment with his Quali Quah powder, on the understanding that if the medicine proved ineffective He Tuck Chaw would lose his head. He Tuck Chaw looked upon these virtual eunuchs, who incidentally included the Emperor's own uncle aged 89, and contemplated his chances of avoiding the headsman's axe.

At this point Lotus Blossom lowered her voice and the audience would hang on every word. 'Gentlemen, in the space of four hours, those horrid examples of lost manhood were flexing muscles unused for decades, and were crying aloud, Pong Wook-ee! which as all of you know is the Chinese equivalent of Eureka! And, gentlemen, there is sufficient quantity of Quali Quah powder in these Vital Sparks, which I am now about to offer, to restore you to the identical condition of virility that has made China a marvel, as well as a problem to the modern world . . .'

The Vital Sparks, which normally retailed at $5 a box, were not going to be sold that evening. No! They were going to be given away with every dollar box of Tiger Fat (there was another story behind this remarkable product, but not quite so far-fetched as the legend of He Tuck Chaw). Mere words probably cannot do justice to the verbal artistry of Lotus Blossom, whose oratory brought inspiration and hope to thousands of men throughout America over a period of thirty years. The audience paid their money, collected their Outer Mongolian medicines, and departed happy. As Holbrook says, it was probably worth a dollar just to hear the 'Pong Wook-ee!' lecture. But do not be misled into believing that this style of sales pitch for an aphrodisiac would not succeed today. It is only necessary to read the story of Pega Palo (page 92) to appreciate that, as recently as 1957, the supposedly sophisticated readership of *Playboy* magazine were persuaded into parting with even more dollars for an equally ineffective remedy from Dominica.

Dr Raphael of Cincinnati with his Cordial Invigorant and Galvanic Love Potions, and Professor Bartholomew with his 'appliance', also made small fortunes in fast time by appealing to the sexually inadequate male. However, the beginning of the end of this golden age came with the successful indictment of T.B. Ganse of the Chicago Cure Company by the US Post Office Department for the fraudulent use of the US mails.[6] It was a strange quirk of American law that enabled a quack doctor to be prosecuted for providing useless remedies by mail order when it was still possible for them to be sold directly to the public.

The Chicago Cure Compnay provided a mail order service for remedies that purported to cure the sexual diseases of men in a permanent and effective manner. Now, although there were virtually no restrictions on the content of advertisements, sections 3929 and 4041 of the US Revised Statutes made it an offence to obtain money through the mails by false or fraudulent means. Following complaints to the authorities by several dissatisfied customers of the company, an indictment was therefore finally brought against Mr Ganse. The particular remedy in question was 'Robusto' and the factor which was principally responsible for the success of the prosecution was the chemical analysis of this wonder cure. Robusto had been marketed successfully since 1894 and consisted of four preparations: red tablets, white tablets, red ellipsoid pills, and round white pills. They were made up by Sharp & Dohme, highly reputable manufacturing chemists who founded what is now a large international pharmaceutical company. The formulae supplied by Sharp & Dohme agreed with those determined by the

Figure 4.2: A Late Nineteenth Century Proprietary Aphrodisiac
Preparation Containing Damiana, Strychnine, Zinc Phosphide and
Cantharides. (This corresponds very closely to the formula for Robusto
described below). The pills would be potentially highly toxic, and a
label on the reverse of the bottle states that they should only be used
on the direction of a physician. The pills themselves have a granular
appearance and are an unpleasant khaki colour

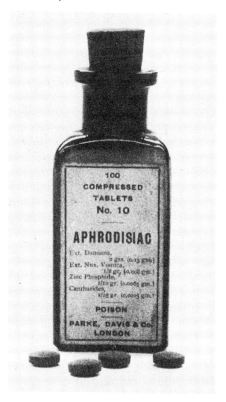

government analyst, but not with those supplied by the defendant.
The actual content, in terms of active ingredients, was as follows. The
red and white tablets were identical apart from their colour and contained extract of nux vomica (strychnine), zinc phosphide (this might
have been zinc phosphite; the former is a highly toxic rat poison),
cantharides (Spanish flies), and damiana; the red ellipsoid pill contained
Blaud's mass and nux vomica. Both these were described as good
general and nerve tonics. The white round pill was compounded of

aloes, extract of podophyllin, nux vomica, belladonna leaves, capsicum, and croton oil, and was described as a mild laxative. None of these formulae would be permitted today, although the concentrations of the active constituents were very low, almost undetectable in fact, and would be unlikely to produce symptoms of poisoning. The illegal aspect of the business centred on the claims made for these products by the Chicago Cure Company. A short sample of their extensive sales literature should be sufficient to indicate the tone:

NEW LIFE!! NEW HOPE!! ROBUSTO: A MAN BUILDER. WHAT ROBUSTO DOES: Robusto is a combination of remedies prepared on a scientific plan so far in advance of the usual stimulating method. It is a system so planned that its powers and effectiveness increase as it progresses, until a cure is established upon a solid basis that will last for life. Those who are looking for a temporary stimulant will please pass us by. We aim to do PERMANENT GOOD, and in no case will we do injury.
ROBUSTO is effective in all cases of WEAKNESS, no matter what the cause. It will BUILD UP any man who is run down, whether from sickness, overwork, dissipation, or whatever else . . . ROBUSTO fills out thin men, gives calm nerves, and imparts self-reliance. ROBUSTO is particularly recommended in all forms of nervous debility and sexual inability. It will stop the life waste attendant upon spermatorrhoea, and restore health to the entire sexual organism.

The advertising matter was written in a similar or even more extravagant style throughout. Specific diseases were named and their cure assured. The success of this campaign can be judged by the fact that, in the year up to 2 June 1904, these remedies had been supplied to between 25 000 and 30 000 clients. The cost of the full course of treatment was $4, so Ganse and his partner must have made a considerable fortune between 1894 and the date of the indictment.

One of the conclusions of law made by the master in chancery in rejecting the appeal against committal is interesting in that it states:

'While a certain degree of exaggeration is justifiable in advertising and extolling the virtues and qualities of that which one is endeavouring to sell, and some degree of commendation and praise of his goods is permissible to a vendor, yet the latitude ordinarily allowed him to puff his wares would not at all justify the extravagant representations and false statements contained in the advertising matter distributed by the complainant.'

The laws have become much stricter since this judgment was made, and any exaggeration is now considered unacceptable.

Secret Remedies

The ability of the authorities to analyse the preparations that were being sold for the treatment of sexual weakness meant that it was possible to determine unequivocally whether or not the remedy could perform the therapeutic miracles which were claimed for it by the manufacturer. The British Medical Association, over the period from about 1904 to just after the First World War, published a series of articles in which the content of many of these proprietary remedies was revealed. The articles also gave the price of the product and the cost of the ingredients. Some of the contemporary 'Nerve Tonics' and their composition are set out below:

Phosferine, 'the greatest of all tonics' and a proven remedy for a whole range of disorders, was sold at 2s 9d for slightly over 1 fl. oz. It contained:

	per cent
Quinine sulphate	0.67
Dilute sulphuric acid	2.5
Dilute phosphoric acid	54.6
Alcohol	8.1
Water	to 100.0

The estimated cost of an ounce of this mixture was ½d.

Guy's Tonic, '. . . a remedy for all the ills to which the flesh is heir' sold at 1s 1½d for 6 fl. oz. and was a relative bargain compared with Phosferine. It contained:

	per cent
Dilute hydrochloric acid	0.59
Dilute phosphoric acid	0.52
Alcohol	2.27
Compound infusion of gentian	40.0
Chloroform water	50.0
Cochineal colouring	a sufficiency
Water	to 100.0

The total cost of these ingredients was calculated to be the same as for Phosferine.

Vitae-Ore consisted of a powder supposedly discovered by 'Professor' Theo Noel, a well-known geologist. The advertising copy spun the following yarn:

> 'I discovered by accident the dried residue of the greatest curative spring in the world. It seemed worth while to try whether this, when redissolved, would have mineral spring virtues. The results astonished me and those who experienced them. The spring was a marvel — far exceeding anything that Homburg, Harrogate, Aix, Bath, or any other spa can show. Some untraceable element defying the chemist's analysis — is it Radium? — enters into the rock . . .'

Vitae-Ore did not present too much of a problem to the chemists who had to analyse it. They reported that it contained:

	per cent
Ferric oxysulphate (anhydrous)	47.57
Magnesium sulphate (anhydrous)	15.89
Water of hydration	36.54

Since these figures add up to 100.00 per cent, there is very little room left for a further element 'defying the chemist's analysis'. The estimated cost of a packet of the powder, for which 4s 6d was charged, was less than ¼d.

There were numerous other nostrums also analysed, including *Cocaphos* and *Damaroids*. In each case they contained quantities of sugar and talc and a dash of quinine, and the profit margins were equally high. One preparation was named *Sequarine* after Dr Brown-Sequard who had carried out the original experiments with testicular extracts (see page 150). There was no evidence that the highly reputable doctor had anything to do with C. Richter and Co. of London who marketed the product. Analysis showed it to contain traces of nitrogenous organic matter, but the nature of this could not be determined. The principal ingredient was alcohol which, at a concentration of 35.8 per cent, made sequarine of equivalent alcoholic strength to neat whisky.

General tonics and cures for sexual weakness were not the only secret remedies investigated by the BMA. It is worth noting that the original Beecham's Powders were subjected to scrutiny and found to contain nothing to justify the claims made for them or the price asked. However, in all fairness to Beecham, who still market these powders as a cold and flu remedy, it should be said that their content has been changed over the years and that the large profits they generate subsidize the costs of research into effective drugs.

Phosphorus as an Aphrodisiac

It will have been noticed that several of the remedies given above contained phosphoric acid. This was not merely a coincidence; there had been several claims made for phosphorus as an aphrodisiac since 1798, and there was increasing evidence that phosphorus, in molecular form, was an essential part of the diet. However, the chemistry of phosphorus is rather complicated and the important differences in properties between phosphorus as a free element and phosphorus in a chemically bound form should be made clear.

The element phosphorus (P) can exist in three allotropic forms, although it does not exist in the free state in nature. Red phosphorus, which is used in the manufacture of match heads and incendiary devices, is not absorbed by the body and hence non-toxic. White phosphorus, in contrast, is readily absorbed and extremely toxic. Black phosphorus is an unusual stable form of phosphorus produced by subjecting white phosphorus to high pressures. The white form was once used widely in the match industry, but has long since been superseded by red phosphorus on account of the highly poisonous nature of the fumes which it gives off, and its tendency to ignite spontaneously on exposure to air. White phosphorus also has the interesting property of glowing in the dark.

The symptoms of acute phosphorus poisoning are nausea, vomiting, bloody diarrhoea, jaundice, abdominal pain, and generalized pruritus or itching. The workers in the phosphorus match industry suffered from an unfortunate disease involving necrosis of the jaw bone, called 'phossy-jaw'. Death would usually occur following a period of delirium and was often due to cardiovascular collapse. Bearing this in mind, how is it possible that phosphorus could ever have been considered to be an aphrodisiac? The solution lies in the confusion between phosphorus as an element and phosphorus as part of a chemical compound. When elements combine to form chemical compounds, the properties of the compound are often dramatically different from the properties of its constituent elements. For example, the properties of sodium chloride or common salt are hardly equivalent to those of the elements from which it is made up. Chlorine is a highly poisonous green gas, and

sodium is a highly reactive soft metal; together, however, they constitute essential mineral components of the diet.

Phosphorus, in the form of its salts, is also an essential element in
the diet, partly because the phosphate salts of calcium $(CA_3(PO_4)_2$,
$CaHPO_4$ and $Ca(H_2PO_4)_2)$ are the only soluble forms of calcium
present in any quantity in the diet, and they are therefore essential
for the absorption of calcium. Secondly, phosphate ions such as
$(PO_4)^{2-}$ are widely used in tissue metabolism as a means of storing
chemical energy. Some vitamins (such as B_6) actually contain phosphorus in chemical combination. The normal diet is very rarely deficient in phosphorus except when the total calorific and protein intake
is severely curtailed. Phosphate salts are occasionally used clinically
in cases where there is a calcium deficiency, in order to increase calcium
absorption.

Phosphate salts and phosphoric acid were once a regular feature of
tonic mixtures, and they still appear in many quack nostrums claimed
to increase stamina and power. The early writers who mention phosphorus rarely make any distinction between the various forms of the
element or its salts, and it is not always clear which form of phosphorus
is being discussed. However, from the descriptions provided of the
effects of phosphorus consumption, it is usually possible to guess
whether the substance in question is red or white phosphorus, or one
of its salts.

Acton, in his book *Functional Disorders of the Reproductive Organs*
published in 1865 (almost two centuries after the discovery of ele-
mental phosphorus) stated that: 'Phosphorus is, in my opinion, one
of those pharmaceutical preparations which the modern surgeon should
most frequently employ in the treatment of impotence.' This reference
is to phosphorus in the form of phosphoric acid, which is non-toxic.
Dr Thompson, in his monograph on the use of phosphorus in medicine,
particularly recommended phosphoric acid in combination with syrup
of oranges and ginger. This recipe sounds almost palatable and would
not be likely to do any harm. However, some of the earlier accounts
of the use of pure phosphorus well illustrate how the element became
thought of as an aphrodisiac.

Alphonse Leroy, writing in the *Magasin Encyclopédique* in 1798,
tells of his own experiences following a dose of 3 grains (about 180 mg)
of phosphorus. He suffered all the classical symptoms of poisoning as
well as unbearable feelings of sexual irritation. A slightly later German
experimenter, Bouttatz, took just 1 grain of phosphorus dissolved in
ether in divided doses over a period of 12 hours. He was more fortunate

and suffered no toxic effects, yet still achieved some degree of venereal excitement.

J.P. Boudet, in his thesis 'Essai sur quelques préparations du phosphore et de ses combinations', published in 1815, wrote: 'Un veillard à qui l'on avait fait prendre quelque gouttes d'éther phosphorée, épouva impérieusement et plusieurs fois le besoin de sacrifier à Venus.' The dose in this case was probably about half a grain. Smaller, medicinal doses are without effect (if any dose of pure phosphorus could possibly be described as medicinal). The short-lasting venereal effect of phosphorus, together with the unpleasant manifestations of poisoning that occur with the larger doses necessary to achieve this effect, led Dr Thompson to conclude: '. . . that phosphorus is not an aphrodisiac unless given in a larger dose than it is safe to employ for medical purposes'.

It is quite possible that the symptoms of sexual irritation have not been uniformly observed in the reports of instances of phosphorus poisoning because they are so transitory. The irritation occurs at an early stage of poisoning and rapidly disappears as further toxic symptoms develop. The minimum lethal dose of white phosphorus in man is thought to be about 1 grain (60 mg), and as little as a quarter of this dose can cause ill-effects. Phosphorus poisoning is no longer as common as it was when it was used in making matches and as a rat poison. It is now classed as a corrosive poison, which can produce a burning sensation in the mouth and throat. This irritation of the soft mucous membranes is quickly transmitted to the bladder and urethra, and this is the most likely basis for its reputed aphrodisiac effects. Irritation of the urethra can lead to prolonged and painful erections. Phosphorus is, in fact, acting in the same way as Spanish flies; both substances are equally poisonous and potentially lethal.

Aphrodisiacs in Early Materiae Medicae

It might be thought that the establishment of quantitative scientific principles during the nineteenth century and the advances in general education would have conspired to diminish belief in the old superstitions, and created a new rational way of thinking. This was not the case as far as sexual function was concerned. The traditional medical practices of cupping, blood-letting and blistering might well have been discarded, to be replaced by new anaesthetics, analgesics and so forth, but the textbooks of materia medica (what we would now call pharmacopoeiae or formularies) still had dark corners. Among the anaesthetics, cardiac stimulants, local anaesthetics and analgesics, there were categories of drug no longer found today. There is little call now for intestinal astringents, purgatives and carminatives, but the aphrodisiacs which appeared owed more to Pliny than to Hippocrates.

Merck's *Manual of the Materia Medica* of 1899, which provided an encyclopaedic list of chemical compounds (and incidentally still does so in more recent editions), included a section specifically entitled 'Aphrodisiacs'. This section has long since disappeared, but in the 1899 edition it included the following: cantharides (used mainly in the treatment of tuberculosis); damiana (an extract from *Turnera microphylla* or *Turnera aphrodisiaca*, which are plants found in the western regions of North America and Mexico that yield a volatile oil and a soft resin; they have the reputation among the natives of western Mexico of being able to stimulate the pelvic organs and, in particular, to restore an atrophied testicle to its former size and function, but according to the National Dispensatory of 1894, there was evidence that damiana could produce haemorrhoids but no good reason to believe that it was an aphrodisiac); gaduol (an alcoholic extract of cod liver oil); glycerinophosphates; gold; muira puama (this does not appear in their main listing); nux vomica (from which strychnine is obtained); phosphorus; spermine (given by injection in cases of debility).

A particularly fine collection of animal products appeared in *Animal Simples Approved for Modern Uses of Cure* by W.T. Fernie MD and published by John Wright & Co. of Bristol in 1899. It was not made clear by the author as to who, apart from himself, had actually approved the remedies he suggests, but for the treatment of sexual disorders in general, and for the strengthening of sexual functions in particular, he lists: ambergris, ant juice, badgers' testes, beaver, civet, cockles, crawfish, cuttlefish, fish in general, lamprey, gold, hart's testicle, hedge sparrow, koumiss, loach, mullet roe, musk, ox marrow, oyster, partridge, pheasant's dung, sheeps' testes, skink, sparrow, sperma hominis, stags' testes, swallow's nest, turkey, viper broth, and viper wine.

Dr Fernie clearly shared with Pliny an exaggerated degree of credulity, but with rather less excuse. No doubt his patients, given the choice, would elect to take the partridge rather than the pheasant's dung. This list does indicate, however, that our emergence from an extended state of ignorance and superstition is perhaps more recent than we would care to believe.

Some of the items listed in the materiae medicae are now recognized as vitamins, which are essential to a healthy diet. In cases where the debility could be ascribed to a deficient diet, the addition of cod liver extract or other vitamin-rich preparation would have proved to be an effective treatment. General 'tonics' of this type are largely irrelevant in western society today, where the diet is unlikely to be deficient in any of the essential vitamins or minerals. Nevertheless, there is still a market for 'tonics' and 'pick-me-ups'.

Before leaving the Victorian pharmacopoeiae it is worth briefly mentioning the anaphrodisiacs they include. Those suggested by T. Lander Brunton in his materia medica of 1893 include some nice examples of the austere medicine applied to patients in need of help in this direction. He includes: ice, conium, camphor, cold baths, potassium or ammonium bromide, potassium iodide, digitalis and, in more general terms, purgatives, nauseants, and bleeding. Dr Brunton also recommends exercise such as rowing or gymnastics rather than walking, which, he suggests, causes the blood to pass towards the lower extremities. Part of this blood may be directed to the pelvic region — and this is to be avoided! Hard mental work and a meagre diet are also suggested as aids to chastity by the good doctor. Some of these measures were used in medieval monasteries as a means of keeping the monks from succumbing to the pleasures of the flesh.

Modern Quackery and Pseudo-science

Gerovital

The discovery that the normal ageing process could apparently be halted or even reversed in patients treated over long periods with injections of the local anaesthetic procaine was made by Dr Ana Aslan of the Bucharest Institute of Geriatrics in the 1950s.[7] She claimed that she had treated tens of thousands of patients over many years who were suffering from the various symptoms of ageing, including cardiovascular disease, arthritis, senility, deafness, neuritis, Parkinson's disease, depression, wrinkled skin, loss of hair, and reduced libido. The procaine was given in the form of a preparation called Gerovital (known also as Vitamin H3, Gerontex, Sex-ex, KH3, etc.), which produced improvements in skin texture, growth of hair and hair pigmentation, muscle strength, mood, and sexual drive. This veritable elixir of life seemed to be offering everything that the ageing patient could desire.

These far-reaching claims for Gerovital were made at a scientific congress held in Bucharest in 1972,[8] and they not surprisingly provoked widespread disbelief among the delegates and a request for some well-controlled clinical trials of the drug. These trials were carried out in several countries with various groups of patients, but they invariably yielded negative results. Although the medical community may not have been able to demonstrate any positive benefit from regular injections of procaine in studies carried out over a period of eight years, the drug is still available in quack preparations for oral dosing, and in a range of rather expensive cosmetics to which Dr Aslan has lent her name. What, then, is the evidence that Gerovital can act as an aphrodisiac?

Pharmacology of Procaine (Novocaine)

Procaine is a local anaesthetic which was introduced as a safer alternative to cocaine as long ago as 1905. It is still widely used and is recognized as one of the least toxic local anaesthetics. It shares none of the stimulant, euphoriant or addictive properties of cocaine, and also, in contrast to cocaine, produces a dilatation of the blood vessels. One problem of using local anaesthetics is that if they are allowed to enter the general circulation, they can produce toxic effects on the heart and brain. However, procaine has the added advantage that it is rapidly broken down by a highly efficient enzyme in the bloodstream. The plasma half-life of procaine is only about 40 seconds, and neither of the two principal metabolites has any pharmacological activity. The plasma destruction of procaine is as shown schematically below:

$$O=COCH_2CH_2N(C_2H_5)_2$$

Procaine

Plasma esterase

Diethylaminoethanol
$$HOCH_2CH_2R$$

$$COOH$$

Para-aminobenzoic acid

This rapid breakdown of procaine is highly relevant to the question of the efficacy of Gerovital. In order to be effective, a drug has to be given in a dose regime which maintains therapeutic levels of the drug at its site of action. The destruction of a drug in the bloodstream can be considered to be complete after five half-lives, so, in the case of procaine, this means that the drug would have been completely destroyed within 4 minutes of an injection. An oral dose of procaine would be even less effective, since a significant proportion of the drug would be broken down in the gut before it even reached the blood-stream.

Formulations of Procaine

One of the reasons suggested by Dr Aslan for the lack of effect of the Anglo-American preparations of procaine was that the original Gerovital was 'buffer stabilized' in a way which prolonged the plasma half-life of the drug. Gerovital for injection has the following formula:

Procaine	100 mg
Benzoic acid	5 mg
Potassium metabisulphite	5 mg
Disodium phosphate	0.5 mg
Water	up to 5 ml

The original dose regime for producing rejuvenation consisted of slow intramuscular injections of 5 ml of 2 per cent procaine solution (the same concentration as above) given daily for 12 days, followed by a 2-week pause. This pattern was continued for a full year, and it required at least four courses of treatment before any beneficial effect could be expected. It was suggested that the benzoic acid formed a stable complex with the procaine, although there is only sufficient benzoic acid to complex 12 per cent of the procaine present. The pH of the mixture has also been invoked as an explanation of the discrepancies, but this factor becomes irrelevant when the mixture is injected since the acid balance of the preparation is rapidly swamped by the near neutral pH of the surrounding tissues. There is therefore little reason to suppose that procaine in preparations of this type behaves any differently from procaine alone.

Many of the other formulations, which are sold either as solutions for injection or capsules for oral dosing, contain a variety of extra ingredients:

Gerioptril (Fischer)	Procaine, vitamins B1, B2, B6, nico-tinamide, sodium pantothenate, inositol, vitamin C, para-aminobenzoic acid
Prokopin 'G' (Woelm)	Procaine, procaine glucoside, bee honey solution, M2 'Woelm', rutin, sodium glutamate
Hormo-Gerobion (Merck)	Procaine, sex hormone, tri-iodothy-ronine, vitamins A, B1, B2, B6, C, D, P, inositol, choline bitartrate
KH3	Procaine hydrochloride (50 mg), haematoporphyrin (0.2 mg), magnesium carbonate (50 mg)

The rather strange formula for Celaton CH_3 Tri-Plus is given in the example of their advertising shown in Figure 4.3. Most of these mixtures are basically providing vitamin supplements that could be obtained more economically by eating cornflakes or yeast extract.

The Rejuvenating Properties of Gerovital

The controversy over the effectiveness of Gerovital, in all its forms, has centred upon the difficulty involved in carrying out a well-controlled clinical trial. When assessing the effects of drugs in geriatric patients, an improvement in their general mood and physical condition is often noted during the period following their admission to hospital, as a direct consequence of the improved care and attention that they are receiving compared with conditions during their previous time at home. Any drug effect must be apparent over and above this background improvement.

Three independent clinical trials of Gerovital were carried out in Britain and the results were published in 1961.[9-11] No improvement was detectable in the patient's appearance, mental state, continence or general mobility. In fact, none of the unpleasant symptoms normally associated with ageing and senility appeared to be alleviated to any significant extent by Gerovital treatment lasting several months. Perhaps most significantly, none of the patients in the study expressed any wish to maintain the treatment after the end of the trial period.

There is an added difficulty in investigating the effects of preparations containing procaine, and that is choosing a suitable placebo. Procaine will produce an obvious local anaesthetic effect around the site of the injection, whereas the placebo should be without effect.

Figure 4.3: A Typical Advertisement for a Preparation Containing,
inter alia, Gerovital. The advertisers have made no specific claim as to
the efficacy of their product, but the sexual connotation has been left
fairly explicit

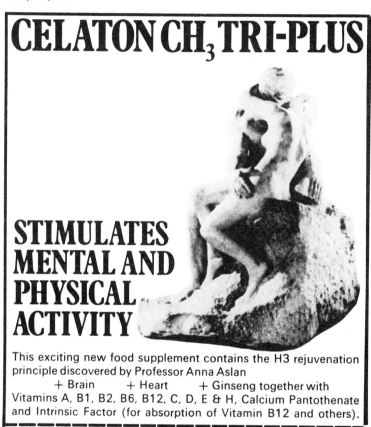

CELATON CH₃ TRI-PLUS

STIMULATES MENTAL AND PHYSICAL ACTIVITY

This exciting new food supplement contains the H3 rejuvenation
principle discovered by Professor Anna Aslan
    + Brain   + Heart   + Ginseng together with
Vitamins A, B1, B2, B6, B12, C, D, E & H, Calcium Pantothenate
and Intrinsic Factor (for absorption of Vitamin B12 and others).

The patient might therefore be able to distinguish between the two
treatments and respond accordingly.

Although Gerovital has been tested clinically for its ability to cure
arthritis, severe senility and relieve depression, there have been no
attempts to assess its reputed aphrodisiac effects. In the light of the
negative findings in all the controlled investigations into its other
properties, this is perhaps not so surprising. Dr Aslan originally claimed
that procaine could stimulate pubic hair growth, increase the level of
circulating hormones, and reverse the physical atrophy of the sexual

organs in both men and women. The improved libido was secondary to
these improvements. The only evidence provided by Dr Aslan was that
chickens treated with a daily dose of 10 mg per kilogram procaine
showed a 9 per cent increase in egg production compared with controls. The relevance of this observation to human patients is not obvious.

The advertisements for procaine preparations in sex magazines make
no mention of the fact that it has only been recommended for geriatric
patients; presumably they would represent too small a proportion of
the magazine's readership. On the positive side there is no evidence
that procaine can do any harm when taken in the doses present in these
multivitamin products, and they at least provide a harmless, if rather
expensive, placebo.

Pega Palo

The fabulous island of Bimini, reputed to be somewhere in the Bahama
group, has long been reckoned to be the source of the legendary 'foun-
tain of youth', whose waters have the power to restore lost vitality,
cure all disease, and promote longevity. The legend was obviously
well known in 1512 when the Conquistador Juan Ponce de Leon set
out to discover this magic source of eternal youth and, presumably,
take advantage of its miraculous properties for himself.

His expedition was unsuccessful, but much later, in 1858, the
commercial potential of such a valuable asset was recognized by a Dr
William M. Raphael of Cincinnati. This entrepreneur of somewhat
dubious background boldly announced that not only had he discovered
the eternal fount, but he had actually drunk from it! He was now prepared to discuss, in complete confidence, the wonderful 'Prolongation
of the Attributes of Manhood' made possible by this fortuitous discovery
at his office at 24 East Fourth Street. The Doctor also proclaimed himself to be an astrologer, philosopher and magician. He was certainly a
fraud and a quack. Nevertheless, his cordial invigorant, at the exorbitant
price of $3 a bottle, sold successfully by mail order over a number of
years.

By the end of the nineteenth century in America the 'Lost Manhood' and 'Secret Diseases of Men' aspect of the nostrum industry was
such a fast-growing section of the US economy that legislation was
introduced in 1906 to curb the greatest excesses of this lucrative trade.
But there is still one fool born every minute, and in the mid 1950s
new generations of gullible individuals hoping to restore their lost
powers were available and ready for exploitation.

The legend of Pega Palo was first reported in *Playboy* magazine, and amounted to the fact that a certain vine, known only to the natives of the Dominican Republic in the Caribbean, could restore virility and had been observed to make men of 80 on the island act like 30-year olds. Visitors to Dominica could obtain the vine in the form of an alcoholic tincture in very small bottles for between $3.50 and $5.00. Its reputation was spread to the USA by tourists returning from the island, and very soon there were several thriving mail order businesses established as far afield as Chicago, Texas and Utah, with the sole object of taking advantage of this highly profitable market. Customers were sold a small piece of the vine with instructions to soak it in rum or other spirit for 8 days before decanting the liquid for use as necessary. Both the Federal Food and Drug Administration and the Post Office Department Fraud Division became interested as well, and it was not long before the first federal seizures of Pega Palo vines took place, the vendors being charged with misbranding (that is, claiming that their product was an aphrodisiac and not providing adequate directions for use on the label). They were also charged with failing to file a New Drug product application. In 1957 the Post Office issued fraud orders against Pega Palo Importers Inc. and several other companies.

One particularly enterprising organization from Texas offered *Tintura de Pega Palo Fortidom* directly to medical practitioners with the following covering letter:[12]

Tintura de Pega Palo Fortidom will not be available for personal use except through physicians until we have completed the investigational stages which we anticipate will be from eight to twelve months.

It is preferred that the Tintura be mixed with a beverage of low alcoholic content such as wine. However it may be mixed with soft drinks, milk, fruit juices etc. It is intended that six tablespoons of the one-to-five mixture will be taken three times daily for the first two days, resulting in the patient absorbing one and one-half tablespoons of Pega Palo the first day and one-half tablespoon the second day. Thereafter he should take a total of six teaspoons three times daily of the one-to-five mixture, resulting in his absorbing a total of three teaspoons of Pega Palo daily. We sincerely appreciate your interest and hope that you will decide to assist us in this project, as we have had a number of inquiries in G——.

Tintura de Pega Palo Fortidom can be furnished to you at $192.00 per gallon, which is $1.50 per ounce, or $50,00 a quart. It has been suggested that the patient be charged $3 per ounce plus the regular charge for an office call, and that the patient be furnished from six to eight ounces at a time. However, this is entirely between the doctor and the patient.

This letter suggests not so much a controlled clinical trial as a highly profitable business venture. In the past, before the advent of the Food and Drug legislation, this would have been a marvellous opportunity for both the manufacturer and the physician to make a great deal of money at the expense of the patient.

The Pega Palo vine belongs to the species *Rhynchosia*, but the distributors of the vine variously described it as *Rhynchosia pyramidalis, R. serpentina, Paullinia sorbillis,* or *P. cupana*. None of these plants is thought to possess any significant biological activity or active principles. The US authorities nevertheless did consider quite seriously the claims made for the vine, and some research was published shortly afterwards in which the effects of the extract of Pega Palo were compared with other alleged aphrodisiacs on various physiological measures in the laboratory rat.[13] The results demonstrated that Pega Palo had no effect on testicular size or weight, nor on the size of the prostate or seminal vesicles in normal or experimentally hypogonadal rats (the Pega Palo was administered to the animals in their normal diet). These particular measures were chosen since the specific claim for Pega Palo was that it restored the normal function of the sex glands in ageing subjects, in whom some testicular atrophy or dysfunction might be expected.

As with the majority of the quack remedies and nostrums of the nineteenth century, the brief reputation of Pega Palo probably owed more to the presence of alcohol in the tincture rather than to any intrinsic activity in the plant extract itself.

Heavy Metals and Arsenic

Arsenic eating is not as uncommon as one might think. At one time it was widely believed that arsenic improved the complexion, increased physical capacity and protected against infectious diseases. It was also popularly held that it acted as an aid to digestion and a sexual stimulant. The basis for this belief, which could be found in France and Germany as well as in England, probably originated in the practice of unscrupulous horse dealers who, by judiciously dosing their worn-out nags with arsensic, could return them temporarily to glossy-coated bright-eyed health. In the regions of Styria and the Tyrol in Austria in the eighteenth century it was not uncommon for country people to eat arsenic regularly as a means of keeping fit and prolonging life.

Arsenic is not such an efficient poison as the writers of crime stories would have us believe. The acute toxic dose of arsenic, in the form of an inorganic salt, is thought to be about 100 mg, but this would not necessarily be fatal since the quantity absorbed from the stomach is very dependent upon the particulate size of the powdered salt. The popularity of arsenic among poisoners may have had something to do with the fact that arsenic was once an ingredient of embalming fluid. The process of embalming would therefore conceal the nature of death and protect the poisoner from discovery. A peculiar feature of arsenic, which is still not completely explained, is that regular doses produce tolerance, so that potentially fatal quantities can eventually be consumed with impunity. The arsenic eaters of the Southern United States (known as dippers) begin by taking about 1 mg in their coffee and then gradually increase the dose until they are consuming perhaps 40 mg a day. This dose would be toxic in most people.

According to Lewin, arsenic was often taken by women and girls in boarding schools, where it was added to the food under medical supervision. This was perhaps intended to improve the complexions of the young ladies; arsenic was certainly popular among the not-so-young ladies of the oldest profession or 'servants of Venus vulgivaga' as Lewin picturesquely calls them. In small doses, arsenic does add colour to the cheeks, partly as a result of damage to the capillaries which allows blood to escape into the surrounding tissues. The other less-pleasant manifestations of arsenic eating include halitosis and flatulence, with a distinct odour of garlic.

Although it is very unlikely that anyone today would knowingly consume arsenic, lead or mercury in the mistaken belief that it had aphrodisiac properties, there have been instances of poisoning by folk remedies that have subsequently been found to contain toxic concentrations of heavy metals. There are several Urdu newspapers published in the UK which carry advertisements for purported aphrodisiacs, and the traditional mixtures usually contain silver or gold in a metallic form. This will not normally be absorbed during its passage through the gastro-intestinal tract, but other toxic heavy metals can be absorbed, particularly if they are in the form of inorganic salts.

A recently published case[14] concerned a 24-year-old male from Bangladesh who had been living in Britain for four years. He was admitted to hospital with severe abdominal pain, nausea and vomiting, and was subsequently found to be suffering from lead poisoning.

Table 4.1: Analysis of Bangladeshi Aphrodisiacs

Sample	Appearance	Lead Content (%)	Other Constituents
1	Brown powder	1.02	Calcium carbonate (chalk) and quartz
2	White powder	46.0	Sodium chloride (common salt) and aluminium powder
3	Pink powder	0.4	Silver (5.78%), iron, zinc, tin, manganese, titanium and strontium
4	Shiny bronze pellets	Nil	Tin sulphide

Eventually he confessed to having taken herbal remedies sent to him from Bangladesh for the treatment of his sexual problems, his main concern being his inability to maintain an erection. These folk remedies obviously contained quantities of lead and so samples of four of them were analysed. The results are shown in Table 4.1.

Quite clearly, none of these so-called remedies would have been the slightest use in helping to maintain an erection, and the second sample would be frankly poisonous. There is no evidence that any of the heavy metals has aphrodisiac or stimulant properties of any kind.

Aphrodisiacs for Adults

The more sexually explicit and even mildly pornographic magazines that are displayed openly at newsagents and bookstands throughout the countries of Western Europe and America represent only a small part of a huge industry involving sex shops, massage parlours, mail-order business, and correspondence clubs. The section of the community for which this particular aspect of modern culture is catering ensures that the market in aphrodisiacs and sex aids represents big business and large profits.

Relatively less legislative control is exercised over advertisements for cosmetics and dietary supplements compared with drugs and medicines, and therefore many of the products offered for sale are promoted as such. A large proportion of the advertisers adopt a somewhat tongue-in-cheek or whimsical approach when describing the potential benefits to be gained from their products, and it is difficult to believe that their readers take them seriously. Nevertheless, the advertisers are careful to avoid making unwarranted claims that could be legally challenged. Currently being offered in UK sex magazines are: 'Motion Lotion (hot and tasty . . . lubricates, invigorates and stimulates at £5.95)'; or even 'Emotion Lotion (both you and she will glow even warmer beneath the silky smooth spell . . . in cherry flavour)'; 'Roman Orgy Love Oil . . . will treble your pleasure'; 'Female Bliss Cream . . . hightens [*sic*] the sensations of the more sensual parts at £3.99 a jar'; 'Sex Sugar . . . it will dramatically increase your sexual urge at £2.99 a box, or £4.99 for two'. The purchase of any two of these products apparently entitles the customer to a free 'Seduction Kit' which would normally retail at £6.95. Surely no one would seriously expect any of these products to make any dramatic difference to their sexual performance.

Moving up market, the advertisements become more subtle and discreet; they are aimed at a more discerning and wealthy clientele but still hint at some vague benefit in terms of sexual attractiveness or performance. The wholly respectable French magazine *Marie-Claire*, as recently as 1980, carried advertisements for 'Galimon . . . (l'aliment de vos nerfs)', a natural dietary stimulant with a broad spectrum of action: '. . . agit sur la fatigue nerveuse, le sommeil, l'humeur, le dynamisme et la mémoire'. Obviously marvellous stuff. A more practical product consists of 'Dermabust . . . a tissue regenerator based on vegetable and animal placental extracts', and with effects that can be imagined from its title.

Within this context it is disappointing to find that the advertising material placed by drug companies in thoroughly reputable medical journals for products of proven efficacy can be as artless and unsubtle as: 'SUSTANON . . . makes a man a man again − in body and in mind'. The product in this case is a standard preparation of testosterone for use in cases of hypogonadism (see Chapter 6). This particular example is regrettably similar to the advertising copy employed by the Chicago Cure Co. at the beginning of this century in promoting their quack cure Robusto.

Adult Aphrodisiacs: What they Contain

A case of accidental 'aphrodisiac' ingestion by two young children in Tennessee in 1978 provided clinicians with the opportunity and stimulus to investigate the exact chemical nature of several products advertised in American adult magazines at the time.[15] Although the preparations described below are fairly innocuous and unlikely to produce any toxic or unpleasant symptoms, this is not necessarily true of illicit aphrodisiacs. A case report in 1970 involved a young man who had been given a capsule of cantharidin as a 'stimulant' at a hippie festival. He subsequently experienced acute poisoning and required hospital treatment.[16]

'Pseudo Spanish Fly' (which fortunately had little in common with authentic cantharidin) was found to consist of ascorbic acid (vitamin C), sugar, and Spanish pepper (capsicum?). Red pepper was also the principal feature of 'Jungle Love' tablets, 'Pseudo-energizer' capsules, and the 'Original Vice-Spice'. 'Pseudo Hard-on Drops' (a liquid preparation) contained liquorice root, ginseng, niacin (a B-group vitamin), and sodium benzoate as a preservative, in addition to the ubiquitous paprika pepper. 'Pseudo Hard-on Pills', in contrast, consisted of white (pepper?), turmeric, sugar cane and ginger. The 'Pseudo Love Stimulant' capsule was 25 per cent garlic, 50 per cent pepper, and 25 per cent sugar cane, a combination unlikely to encourage oral intimacy.

The 'pseudo' aphrodisiacs promoted by adult magazines appear to be a non-serious attempt to provide sexual titillation by inference and suggestion rather than pharmacologically active products. For people in real need of drug therapy or sexual counselling for their sexual problems, there are far more effective sources and means of treatment through recognized medical practice (see Chapter 6).

5 THE CLASSICAL APHRODISIACS

If you were to ask anyone today if they could think of an aphrodisiac, they would be almost certain to suggest rhinoceros horn or even Spanish flies as examples. A medievalist or other historian might also manage to recall the magic mandrake, whereas a member of the contemporary Californian drug scene would certainly be able to suggest Mandrax, a widely available drug once used as a sleeping pill and coincidentally named by the manufacturers to suggest the classic drug. On the other hand, very few people are likely ever to have used rhino horn, Spanish flies or mandrake for aphrodisiac purposes, although they might well have casually noticed the disinhibiting effects of alcohol on social and sexual behaviour.

The persisting reputations of rhinoceros horn, Spanish flies and the mandrake deserve more detailed examination in order to assess whether or not they can be justified. At the same time, there is no doubt that alcohol is currently in popular use as a practical social aphrodisiac in all but name. This chapter is therefore devoted to a review of the reality behind the widespread historic belief in the efficacy of these four substances.

Rhinoceros Horn

The value of the horn of the rhinoceros as a sovereign cure-all can be traced back to prehistory. The animal depicted in palaeolithic cave paintings was the Asian or single-horned rhinoceros; in other words, the unicorn. The concept of the unicorn as an elegant equine creature with a horn on its forehead is a far more recent invention. It is almost certainly the single-horned rhinoceros which is responsible for biblical references to the unicorn. For example, in Job 39.10:

'Canst thou bind the unicorn with his band in the furrow,
Wilt thou trust him because his strength is great?'

This certainly sounds more like a heavy lumbering beast than an animal capable of cantering or galloping in carefree abandon as pictured on heraldic devices. Nevertheless, there is a long-standing tradition of magic associated with the unicorn, whose horn was much prized no matter that the beast itself was a myth.

Ctesias, the Greek historian, gives the earliest description of the unicorn in an account written about 400 BC. It was about the size of a horse, he says, with a white body, a red head, blue eyes, with a horn a cubit (50 cm) long on its forehead. Towards the base the horn was white, black in the middle part, and red towards the extremity. The animal was very swift and strong, and, though not fierce by nature, it fought desperately when attacked, using horn, feet and teeth, so that it was not possible to take it alive. Drinking cups purportedly made from the horn would preserve the drinker from spasms, epilepsy and poisoning. In order to catch a unicorn to obtain the material necessary for the manufacture of such a valuable drinking vessel, medieval legend has it that a virgin was to be placed near his haunt (the unicorn was invariably male, of course). The beast would use his keen scent to find her and then lay his head upon her lap and fall asleep. Professional hunters apparently used even more devious means to ensnare this innocent creature:

'For there is in their nature, a certaine savor, wherewithall the unicornes are allured and delighted; for which occasion the Indian and Ethiopian hunters use this stratagem to take the beast. They take a goodly, strong and beautiful young man, whom they dresse in the apparell of a woman, besetting him with divers or odoriferous flowers and spices. The man so adorned they set in the mountains or woods where the Unicorn hunteth, so as the wind may carrie the savour of the beast. The Unicorne deceaved with the outward shape of a woman, and sweete smells, cometh to the young man without feare, and so suffereth his head to bee covered and wrapped within his large sleeves, never stirring, but lying still and asleep. Then the hunters come upon him and, by force, cut off his horne and send him away alive.'

John Guillam, the author of *A Display of Heraldrie*, wrote in 1610, lest anyone should consider this to be a fairy tale and too far-fetched to be true, that: 'Some hath made Doubt whether there be any such Beast as this, or no. But the great esteem of this Horn (in many places to be seen) may take away that needless scruple . . .' 'This great esteem' seems to have centred on the reputed antidotal properties of the horn against the bites of venomous animals. This is an Asian tradition derived possibly from early Chinese medicine; it is not mentioned by Greek writers on the subject. This specific application of rhinoceros horn is still found in Asia. The horn seen 'in many places' was undoubtedly the horn of the narwhal, which possesses a straight horn with a spiral twist. The discovery of the living narwhal by European seamen exploring the North Atlantic and Arctic oceans in the latter half of the seventeenth century provided the explanation for these strange horns that were occasionally found washed up on beaches. The value of the horn, which had once been greater than gold on an equivalent weight basis, declined rapidly as the truth became known. It was eventually removed from the British Pharmacopoeia in 1742. A particularly fine example of a carved narwhal horn dating from this period can still be seen in Chester Cathedral.

Although the myth of the unicorn had been discounted by the middle of the eighteenth century in Europe, the worldwide belief in the sexual efficacy of the horn has unaccountably persisted in spite of the fact that most of the contemporary material is obtained from the heavily poached black or grey African rhinoceros which possesses two horns. The various preparations derived from this horn are used in Chinese and Korean aphrodisiac potions as well as in antidotes to poison. There are many anecdotal accounts by explorers and anthropologists travelling in Asia of the success of such preparations in cases where all other treatments had failed. The most obvious explanation for the aphrodisiac reputation of the horn lies in homoeopathic magic; it takes little imagination to relate the shape of the rhino horn to the erect penis (see Figure 5.1).

Is it then possible that the horn actually contains some active constituent? The horn consists of hard, fibrous tissue which is chemically and mechanically similar to hair. In common with wool and hoof tissue, it contains a high proportion of cross-linked proteins or keratins. It is the sulphur atoms in these proteins that contribute to the unpleasant smell associated with the burning of keratinous tissues. Extracts of rhino horn have been analysed and found to contain, like any similar tissue structure, quantities of calcium and phosphorus; the latter, of course, has always had a reputation as an aphrodisiac (see page 83). A diet deficient in either of these essential elements might result in muscle weakness and general lassitude, and the addition of a phosphorus-rich food under these circumstances would be quite likely to improve physical vigour and perhaps even rekindle an interest in sexual activity.

The addition of phosphorus or calcium to an otherwise wholly adequate diet will have absolutely no effect on physical performance

Figure 5.1: The Single-horned Rhinoceros. Based on an illustration in *The Life of Vertebrates* by J.Z. Young

except from a psychological point of view. Also, the very high cost of preparations of rhino horn probably means that the quantity consumed would be so small as to make a negligible contribution to the overall dietary intake of either element. The only active principle in a preparation of rhino horn seems to be its reputation. It is unfortunate that an ancient and thoroughly discredited myth should be in a great part responsible for the illegal slaughter and possible eventual extinction of the African rhinoceros in the wild.

Spanish Flies

Spanish flies or cantharides are the common names for a variety of blister beetles of the *Cantharis* or *Mylabris* genus as well as *Lytta vesicatoria*, which are found commonly in mediterranean countries, the Near East, and even in Britain. The entire beetles are dried and crushed to a powder, which then contains between 0.4 and 1 per cent of the active principle, cantharidin (hexahydro-3,7-dimethyl-4,4-epoxyisobenzofuran-1,3-dione). Pure cantharidin can be extracted from the beetles and crystallized as the lactone of cantharidic acid. It has the following chemical structure:

In the past cantharidin has featured in formularies either as cantharis (the dried powdered beetles) or as a tincture or alcoholic extract of cantharidin. Its use was principally as a blistering agent or vesicant, and in the British Formulary it was classified as a Schedule I Poison. It last appeared in the British Pharmacopoeia in 1953, but it still lingers on in veterinary medicine as a constituent of blistering fluid and in some proprietary wart-removers. The toxic dose in humans can be as little as 3 mg, or one ten-thousandth of an ounce; the fatal dose is only 32 mg. If it is taken orally, the first symptoms consist of an intense burning sensation in the mouth and throat, severe abdominal cramp and vomiting, followed by diarrhoea and the passing of blood in the urine. Over a period of time, urination becomes increasingly difficult and painful. The victim usually succumbs to kidney failure and shock within 24 hours of taking the poison. A very vivid account of the effects of Spanish flies appears in Fernie's *Animal Simples* of 1899:

> When swallowed, the Spanish Fly acts as a violent acrid poison, exciting angry inflammation within the stomach and bowels, affecting also the nervous system profoundly, and setting up ardent pain in the bladder, with difficult bloody urine, and straining. The throat becomes constricted and swallowing is difficult; griping and great tenderness of the belly are caused, with giddiness; in severe cases convulsions, delirium, insensibility and death.

With this in mind, it is very difficult to imagine how cantharides could possibly have gained such an unequalled reputation as an aphrodisiac; indeed, the very name Spanish fly has become almost synonymous with sexual stimulation. In *Roget's Thesaurus*, for example, under the heading 'stimulant' can be found: 'aphrodisiac, philtre, love philtre, cantharides, Spanish fly.' By turning in the same

Figure 5.2: *Lytta vesicatoria*, One of the Large Number of Beetles that Contain the Blistering Agent, Cantharidin. This wide range of species is known generally as Spanish flies, and they are usually brightly coloured

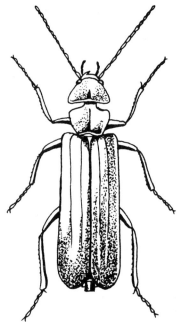

book to the heading 'be in love' one finds 'to burn, sweat, faint, burn with love, glow with ardour, flame with passion', and so on. Love and passion can certainly produce some very strong physiological effects upon the body, but we clearly recognize these poetic metaphors for what they are rather than the actual physical conflagration of the body. Yet, knowing how the Law of Similarity can be applied, this is very possibly where some of the myth has arisen.

In a romantic and less scientific age it was probably very easy and natural to equate the emotional or psychological effect of love and desire with the physical effects of an artificial stimulant. In other words, if love or desire evokes burning, fainting and distraction, and cantharides produce the same symptoms, then may not cantharides also provoke love and desire? This type of false logic, which is the basis of all sympathetic magic, would not have seemed so unreasonable in a less enlightened age.

It is also possible that the irritation to the urinogenital tract produced by cantharides could lead to an urge for relief that might be achieved through sexual intercourse. Some of the older references to cantharides mention cases of persistent erection or priapism, which may be extremely painful, particularly if urination becomes impossible.[1] None of the more recent cases in the literature makes any mention of increased sexual appetite or urgency.

Tacitus, in his Annals of Imperial Rome, mentions that cantharides were available and probably used by the Emperor Tiberius's mother, Livia, to stimulate the sexual appetites of other members of the imperial family in order to encourage them into committing sexual indiscretions which could later be used against them. The successful long-term use of cantharides in this respect seems most unlikely. The Marquis de Sade used cantharides and, in one instance, is supposed to have poisoned the inmates of a brothel to the extent that several of the women committed suicide by throwing themselves out of the windows, such was the intolerable pain produced by the drug.[2] Madame de Pompadour apparently used tincture of cantharides to regain and maintain the sexual attentions of King Louis XV.

There are numerous references in the classical erotic literature of both Europe and Asia to the use of drugs having effects that correspond to those of cantharidin. However, the mildness of the symptoms as described can be very misleading. For example the following passage from the erotic English classic, *Fanny Hill*, by John Cleland, provides an interesting, if fictional, account in which the possible use of Spanish flies may be inferred:

'Accordingly he went out and left me, when a minute or two after, before I could recover myself into any composure for thinking, the maid came in with her mistress's service, and a small silver porringer of what she called a bridal posset, and desir'd me to eat it as I went to bed, which consequently I did, and felt immediately a heat, a fire run like a hue-and-cry thro' every part of my body; I burnt, I glow'd and wanted even a little of wishing for any man.'

Such a potent recipe today would ensure a fortune for its possessor, but it is highly unlikely to have contained any cantharidin. Most of the symptoms described by Fanny Hill could be produced by a neat spirit which would act almost immediately. Cantharides take much longer to act, and a very low dose might be expected to have similar effects to chilli peppers. There has never been any reason to suggest that curried dishes have any aphrodisiac effects.

A more explicit reference to the use of Spanish flies appears in John Gay's *Beggar's Opera*:

'Strait to the pothecary's shop I went and in love powder all my money spent behap what will, next Sunday after prayers when to the ale house Lubberkin repairs. Then flies into his mug I'll throw and soon the swain with fervent love shall glow.'

Cantharides in Allopathic Medicine

The medicinal application of cantharides reached a peak when they were used particularly for the production of blisters in the treatment of pleurisy and pericarditis. Orthodox physicians of the eighteenth and nineteenth centuries were known as allopaths (from the Greek for 'other') because they treated illnesses with counter-irritations to induce some new action that would be incompatible with the original disease. This is the exact reverse of homoeopathic medicine which is based on the principle that like will cure like (see below). Allopathic treatments included cautery (burning), bloodletting (either by surgical incisions, venesection, or the application of leeches), or vesication (blistering).

In the case of pleurisy, the rationale for the unlikely treatment involving blistering was that the excess fluid in the lungs or on the chest would be drawn into the heroic blisters raised on the skin by the topical application of cantharidin, usually in the form of plasters. The pharmacopoeia of the King's and Queen's College of Physicians in Ireland in 1850 gives the following recipe for tincture cantharidis: 'Take of Spanish Flies, in coarse powder, half an ounce; Proof Spirit, two pints; Macerate for fourteen days, strain, express, and filter.' The practical applications of such a mixture used in the form of a plaster are well described in a medical textbook of the same period:

With a view to their general stimulation, they are used in low states of disease, requiring an excitant influence to support the action of the heart, and those of the nervous system. Their first effect is to inflame the part to which they are ·applied; and the subsequent vesication [blister-forming] is the result of the inflammation, the distended vessels relieving themselves by effusion . . . epispastics [the alternative term for vesicants] are, therefore, primarily local and secondarily general stimulants.

But when the depression is sudden, arising from some sedative but temporary influence on the nervous system or circulation, or when, if more protracted, it is sustained by a similar influence more persistent but still of limited duration, there is an indication for these with other stimulant remedies, in order to arouse the great functions from the torpor in which they may have been left, or to sustain them until the depressing agency shall cease.

Having expounded on the indications for this particular treatment, the appropriate 'dose' is then discussed:

The size of the blister is of considerable importance. As a general rule it should be large; as the pain and inconvenience are little increased, while the remedial impression is proportioned to the extent of the blistered surface.

Apparently, blisters of 12 in by 6 in (30 cm by 15 cm) were not uncommon in those days. Today, this highly unscientific use of a corrosive poison in what was then traditional medical practice seems terrifying, although a milder vesicant or rubefacient is still available and used today in the form of the mustard plaster. At the same time as blistering was being propounded in the treatment of pleurisy, it is worth noting that lead acetate was recommended for reducing inflammation, and that mercury and its salts were the first line of treatment for venereal disease. In fact, many of the early reported symptoms of syphilis were probably symptoms of mercury poisoning. The spurious reputation of cantharides as an aphrodisiac was maintained throughout a period when knowledge of the mechanism of action of drugs in the treatment of disease was virtually non-existent. However, whereas contemporary medical science has since advanced to the point of making the normal clinical practices of a century ago seem medieval in their primitiveness, the aphrodisiac reputation of Spanish flies seems to have survived untarnished. For a brief period cantharidin was included in commercial aphrodisiac preparations (see Figure 4.2).

The contemporary use of cantharidin is restricted to a few preparations for the removal of warts, but even now this can have hazardous consequences, as shown by the following report of a case of self-poisoning in Canada.[3] A girl of 18, after an argument with her boyfriend, swallowed between 1.5 and 2 ml of a wart-removing fluid (Cantharone) which contained 0.7 per cent cantharidin. She suffered from ulcerations of the mouth, pharynx and oesophagus. When she was admitted to hospital, her blood pressure had fallen and an ECG examination of her heart indicated that some damage to the heart had occurred. However, she recovered over a period of ten days and there were fortunately no long-lasting ill-effects.

Cantharidin has long been used in traditional Chinese medicine as an antitumour agent, and there is much research being carried out to assess its potential use in the treatment of cancers. Animal experiments reported in the Chinese scientific literature[4] suggest that it may be effective, although the therapeutic dose is apparently very close to the toxic dose.

Cantharides in Homoeopathic Medicine

Homoeopathic medicine is attracting an increasing amount of interest, particularly among people who do not respond to conventional treatments. The principles on which homoeopathy are based are that 'like will cure like', and that a complete investigation of the patient should be made at the time of consultation so that the diagnosis is not based merely on one or two superficial symptoms. The remedies which are then prescribed are administered in such small dilutions that, to the scientist, it seems that any beneficial effect must be due to auto-suggestion or a placebo response (see Chapter 6). However, homoeo-paths believe that the drug initially imparts some intrinsic property to the material in which it is dissolved and that this property is maintained even when the drug has been diluted to the point at which it is doubtful whether a single molecule remains. This does sound rather like the principle of the Law of Contagion, which was dismissed in Chapter 3. However, this is only one aspect of homoeopathic practice.

In order to get some idea of the dilutions involved in the preparation of homoeopathic remedies, it is worth taking the example of cantharidin. It is used in the form of cantharis which is made up in a solution of one part in a hundred. One part of this solution is again diluted 100 times, and this procedure is repeated until thirty sequential dilutions have been completed. This is a 30c potency and represents, in scientific terms, a dilution to 10^{-60} of the concentration of the original.

The cantharis prepared in this way is used for the treatment of painful symptoms in the bladder and urethra (e.g. cystitis), for relieving erysipelas, gnat bites and the blisters from burns and scalds. In other words, it is used for treating all the symptoms which cantharides, at a higher dose, would produce. It would be most interesting to know whether patients taking this treatment have ever reported any untoward aphrodisiac side-effects.

Some Instances of Cantharidin Poisoning

One peculiar case which highlights the dangers of using cantharides in any context, was reported in the *British Medical Journal* in 1954.[5] A very keen fisherman had obtained a small quantity of cantharidin from an illicit source with the object of mixing a bait with something 'sexy' in it in order to attract the fish more readily. He mixed the cantharidin with water in a bottle and, putting his thumb over the end of the bottle, proceeded to shake the contents vigorously. Unfortunately, cantharidin is only very slightly soluble in water, and it was doubly unfortunate for the fisherman that he did not have a stopper for the bottle. Some undissolved cantharidin was left on his thumb and shortly afterwards he accidentally pricked the same thumb with one of his fishhooks. He naturally sucked his thumb and within an hour began to feel sick. He very quickly developed all the classical symptoms of cantharidin poisoning and, in spite of all the efforts of the local hospital, died six hours later.

An earlier and less tragic history of a case of poisoning was reported by Dr J. Meynier in 1893.[6] This case also had some most unusual features. The author was the medical officer to several battalions of French troops sent to construct a new outpost in Sidi-bel-Abbes in French North Africa in the April of 1869. During the following month an increasing number of soldiers began to report sick with unusual but distinctive symptoms. In the words of the doctor:

'Les malades se plaigneraient de sécheresse de la bouche et de soif; le région stomacale était douloureuse; l'appétit, excité chez les uns, était diminué ou même aboli chez les autres. Ils accusaient aussi de fréquentes envies d'uriner; la miction était douloureuse, bien que les urines fussent plus abondantes; il y avait de l'ardeur du col vesical et de la vessie; quelques-uns ont présenté du priapisme ou du moins les érections douloureuses et prolongées de la chaudepisse cordée.'

The symptoms of dry mouth, gastric pain, and painful and difficult passing of urine we would now recognize as being strongly suggestive of cantharidin poisoning. The patients also complained of lassitude, faintness and chills. Dr Meynier, presumably not being familiar with this particular condition, took account of the region of the body most obviously affected and the established reputation of his patients and made an initial diagnosis of syphilis. However, since it had been several months since the men had been anywhere near a woman, he had to dismiss this possibility. Eventually, having eliminated the possibility of food poisoning arising from the camp kitchens, he found the common factor which linked all the cases. The men in question had been catching and eating the local frogs which apparently abounded in the Melzera river which ran close by the camp. The frogs themselves were not poisonous, but a period of observation of their behaviour at the river soon revealed that they were feeding on a large population of beetles which regularly fell into the river from the overhanging trees. The beetles were easily identified as belonging to the *Mylabris* or blister beetle genus. The doctor referred to them as Greek flies.

There are several interesting points arising from this account; first, that it is one of the few well-documented accounts that mention prolonged erections as a consequence of consuming cantharidin. The other symptoms which these patients also experienced suggest, however, that they gained little or no pleasure from their condition. The second point is that the frogs were able to consume the beetles with impunity, even though the cantharidin was still present intact and was clearly not broken down by the metabolic processes of the frogs. Finally, it may be worth remarking that perhaps only a regiment of French soldiers could have suffered from such an unusual source of poisoning.

There have been several instances of homicidal poisoning by intent, in which young girls have been given cantharidin disguised in either drink or sweets. One particular case occurred in London in 1954,[7] coincidentally in the same year that the over-zealous fisherman poisoned himself. Two female clerks aged 19 and 27 were given some coconut ice by a male employee at the same company where they worked. Unknown to them he had filled the sweets with cantharidin. Both girls died and the man responsible was charged with two counts of manslaughter. He pleaded guilty and received five years' imprisonment.

Cantharides obtained from a variety of insect sources are used by herbalists and medicine men in various parts of the world either as a cure for venereal disease or as an abortifacient. In South Africa, for example, the incidence of cases of cantharidin poisoning officially reported reached a peak of about 20 cases a year during the late 1930s. It is not clear what proportion of these cases represents intentional rather than accidental poisoning.

A letter to the *Pharmaceutical Journal of Great Britain* in 1910 suggests that weak tinctures of cantharides were, at that time, one of the standard cures for baldness, the effectiveness of the treatment being vouchsafed by the sale of several hundred bottles of Squire's

Linimentum Crinale. The author of the letter proposed a slight modification to Squire's original recipe of 1890 by the addition of potash soap to the mixture. This apparently improved the penetration of the mixture into the tissues and thus rendered it more effective. It would be interesting to know whether the large sales of this hair restorer reflected its use in another direction by the ageing male customers who presumably purchased it.

From all the evidence presented, it should be evident that Spanish fly possesses no aphrodisiac properties but is, in fact, a potent and highly dangerous corrosive poison. Its ethical use has declined to the point at which it is no longer widely available and its illicit use is much less likely.

In conclusion, it is interesting to consider why various species of beetle should contain such an unpleasant substance. The answer is probably that it makes the beetles extremely distasteful or even toxic to a potential predator and so protects them, although in the case of the North African frogs mentioned earlier, this did not seem to be an adequate defence. Cantharidin is certainly best left to the beetles, in which it presumably does serve some useful purpose.

The Mandrake

The legend of the mystical mandrake can be found in Greek, Roman, Arabic, Indian and Chinese writings. It seems to have been a universal myth in most cultures. It is mentioned in the Old Testament and in Anglo-Saxon and medieval writings. It is frequently referred to by Shakespeare and the other Elizabethan dramatists, but the end of that century marked its rapid decline into obscurity. The mandrake as a magic element of alchemy and sorcery may have disappeared, but vestiges of the legend have persisted in the quack remedies marketed at the turn of the present century (see Figure 5.3). Its most recent revival came about with the launching of a new sleeping pill named Mandrax in the mid-1960s. Although this was only one of many trade names used by the pharmaceutical companies for the parent drug, methaqualone, it was perhaps one of the factors that contributed to the epidemic of abuse of the drug for sexual purposes that followed, and which eventually prompted its withdrawal from clinical use.

Before embarking on the long history of the mandrake, it is important to establish the distinction between the magic mandrake, an entirely mythical plant used in sorcery, and the real mandrake, which is

Figure 5.3: Joshua Barrett's Mandrake Embrocation. This was not sold as an aphrodisiac, but the label includes an amusing visual pun

Source: History of Medicine (1971), 3 (1), p. 20

represented by several unrelated species possessing similar hypnotic and sedative properties by virtue of the alkaloids they contain.

The Magic Mandrake

This remarkable plant was supposed to grow only at the foot of a gibbet and to glow in the dark. The active part of the plant was the

Figure 5.4: The Recommended Anglo-Saxon Technique for Extracting the Mandrake Root from the Ground by Use of a Trained Dog. The drawing is based on an illustration in an early English herbal. The anthropomorphism of the plant has been greatly exaggerated by the early artist. A more representational version of *Mandragora officinarum* is shown in Figure 4.1 (b)

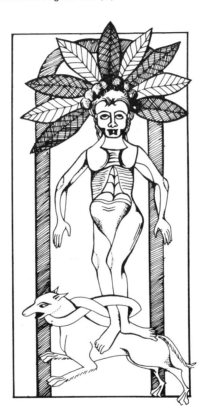

root, which could resemble the figure of a man. The closer this resemblance, the more powerful the root. As we shall see later, many less than scrupulous witches were not above doctoring their mandrake roots with a little judicious carving in order to improve the appearance and the value of their specimens. The root was an expensive ingredient in a charm since its collection required a procedure which was fraught with danger. On pulling the plant from the ground it was supposed to give vent to a shriek that was instantaneously fatal to anyone within earshot.

To overcome this difficulty, the collectors of the mandrake tied a dog to the plant and, after retiring to a safe distance, would call the animal. It has not been recorded whether the dog suffered any ill-effects from this exercise, but an illustration from an early herbal clearly shows this technique in practice (see Figure 5.4).

Over the years, each successive writer added another layer of fable to the story of the mandrake. In the herbal of Bartholomaeus Anglicus of about 1260, for example, the standard legend is recorded with some extra details: 'They that dygge mandragora be best to beware of contrary wyndes whyle they digge. And maken circles abowte with a swerder and abyde with the dyggynge until the sonne goynge downe.'

Once the precious root had been obtained in this way, the next step was to prepare it for consecration. Thomas Lupton, writing in his *A Thousand Notable Things*, gives an account of such a procedure:

> Take the great double root of briony [one of the plants reputed to be the mandrake] newly taken out of the ground, and with a fine sharp knife, frame the shape of a man or woman, with his stones and cods and other members thereto, and when it is clean done, prick all these places with a sharp steel, as the head, the eyebrows, the chin and privities, and put into said holes the seeds of millet or any other that brings forth small roots that do resemble hairs . . . after this put it into the ground and let it be covered with earth until it hath gotten upon it a certain little skin, and then thou shalt see a monstrous idol and hairy, which will become the parts if it be workmanlike or cunningly made or figured.

This sly practice would normally take place at a full moon. The root would be dried and perhaps at some later witches' ceremony be consecrated and used in the preparation of a love charm or even as the vessel for a familiar spirit. Such activities were considered dubious by the herbalists of the time. For instance, in the *Grete Herball* of 1485:

> Some say that the male hath figure of shape of a man. And the female of a woman, but that is fals. For nature never gave forme or shape of mankynde to an herbe. But it is troughe that some hath shaped suche fygures by craft as we have fortyme herde say of labourers in the fields.

This is the earliest record of a writer actually questioning the validity of the myth; later herbals are even more forthright in their condemnation of the practice:

The rootes which are counterfeited and made like little puppettes and mammettes which come to be sold in England in boxes with heir and suche forme as man hath, are nothyng elles but folishe fened trifles and not naturall. For they are so trymmed of crafty theves to mocke the poore people withall and to rob them both of their wit and theyr money.

Turner's herbal of 1551 makes clear that: 'It groweth not under gallows as a certain doting doctor of colon in his physick lecture dyd teach hys auditors.'

The final denouncement of the magic mandrake was made by Gerard in his herbal published in 1597:

There have been many ridiculous tales brought up of this plant, whether of old wives or runnegate surgeons, or physick mongers I know not, all whiche dreames and old wives tales you shall hence-forth cast out of your bookes of memorie.

The Plants known as Mandrake

Many of the early accounts that refer to the mandrake are sufficiently precise to suggest that they are referring to a specific plant:

The mandrakes give a smell, and at our gates are all manner of pleasant fruits, new and old, which I have laid up for thee, O my beloved.

Song of Solomon 7.13

Early Arabic writings describe the mandrake as having small round fruits, orange-red in colour and known as *baid el jinn* (eggs of the djinn). Mandrakes also feature early in the Bible, in the book of Genesis (xxx. 14-17). where it is related how Jacob's son Reuben found some mandrakes in the field and gave them to his mother, Leah. In return for these mandrakes, Rachel, who was Jacob's first wife, allowed Leah to sleep with Jacob for that night. The implication is that the mandrake had some use either as an aphrodisiac or as a treatment for infertility.

In early European and Middle-Eastern writings, the plant named as the mandrake is almost certainly one of the Solanaceae (potato) family and quite possibly the thorn-apple (*Datura stramonium*), known in the United States as Jimson Weed. In some texts this plant is called mandragora, although another plant, *Mandragora officinarum*, does exist. The roots of this plant are similar in external appearance to those of ginseng (see Figure 4.1a), and this similarity may contribute to the reputation of ginseng as a panacea. Thorn-apple contains several alkaloids, notably *l*-hyoscine (scopolamine), which is pharmacologically similar in its properties to atropine. Extracts of *Mandragora officinarum* have yielded *l*-hyoscyamine, *l*-scopolamine, and another alkaloid called mandragorine. It is therefore evident as to how *Mandragora* and *Datura* could have been called mandrake and thus have been confused. Both plants contain alkaloids with similar properties and would have similar effects if eaten. *Datura* was used in the so-called 'flying ointment' prepared by witches to enable them to fly to their sabbaths.

Chinese references to the mandrake are certainly describing ginseng. When sold today, ginseng is often described as the 'man root', and the illustrations on the packaging tend to employ artistic licence to the extent that it is the mythical mandrake which is portrayed rather than the root of *Panax ginseng*. (The aphrodisiac potential of ginseng has been assessed in Chapter 4.)

The English Mandrake

The plant which is occasionally, and wrongly, described as the mandrake in the English herbals is white bryony (*Bryonia dioica* Jacquin), a member of the cucumber family found growing wild, particularly in southern Britain. It is poisonous and contains, among other things, the glycoside bryonin, which is an irritant purgative rather than a sedative or hypnotic. However, the large fleshy tuberous root can resemble the shape of the mandrake root. In France it is known as Devil's Turnip, and cases of poisoning have occasionally occurred as a consequence of either mistaking the root for an edible parsnip, or by intentional consumption as a folk remedy for reducing lactation in women. It apparently has a very acrid taste which renders it most unpalatable.

The black byrony (*Tamus communis*) is botanically unrelated, being a member of the yam family. It also has large fleshy roots, which are somewhat darker than those of white bryony. The active principle has not been identified with certainty, although it appears to be similar to bryonin in its actions.

The American Mandrake

The American mandrake is something quite different. It is *Podophyllum peltatum*, also known as may apple, wild jalap, wild lemon, devil's apple, umbrella plant, vegetable mercury, yellow berry, etc. It is unrelated to the European plant either botanically or in terms of its active constituents. The unripe fruits, rhizome and roots contain up to 6 per cent of podophyllum, and also podophyllotoxin, α-peltatin and β-peltatin. The latter are responsible for the powerful purgative properties of the plant, and podophyllum is currently used in a tincture to dissolve warts. May apples, in spite of their name, are not edible and have no reputation as aphrodisiacs.

Mandragora

Early references to the effects of mandragora make it clear that the drug had soporific and sedative properties. In Greek mythology Circe used a potion containing mandragora to help her seduce Odysseus's men. In some places mandragora is referred to as the 'Herb of Circe'. In the original Greek legend, the god Hermes fortunately warned Odysseus of the fate that was in store for him and told him that the effects of Circe's potion could be reversed by the use of garlic. Smelling salts might be the modern preferred means of bringing someone back to their senses, but garlic has been used until relatively recently as a restorative. Sidney Smith, writing in 1813, said: 'I have been spending some weeks of dissipation in London, and was transformed by Circe's cup not into a brute but a bean. I am now eating the herb moly in the country.'

Mandragora was a very popular pain-killer and soporific throughout the Middle Ages. The soporific sponge described in the *Antidotarium* of Nicholas of Salerno, published in 1471, contained a mixture of mandragora, hemlock, opium, henbane, mulberry juice, lettuce and dried ivy. There is no doubt that the inhalation of this potent narcotic mixture would have had soporific effects.

The use of mandragora as an aid to sleep and forgetfulness was obviously well known in Shakespeare's time:

'Not poppy, nor mandragora,
Nor all the drowsy syrups of the world,
Shall ever medicine to thee that sweet sleep
Which thou ow'dst yesterday.'

Othello III.iii

'Give me to drink mandragora.
That I might sleep out this great gap of time
My Anthony is away.'

Anthony and Cleopatra I.v

In the twentieth century, general anaesthesia for major surgical operations is normally preceded by the administration of a pre-medication by the anaesthetist. For a long time the drugs most commonly used have included morphine derivatives (for relief of pain and anxiety) and scopolamine (hyoscine), which relaxes the smooth muscles, dries up the bronchial secretions and saliva, and also has sedative effects. In combination they induce a pleasant feeling of well-being in the patient and often a period of total amnesia. Thus, modern clinical practice is taking advantage of the properties of plants that were recognized centuries ago. The only difference is that modern drugs are of established purity and potency and their mechanism of action is understood.

The behavioural effects of mandragora are likely to be very similar to those of the other sedative hypnotics and general depressant drugs such as alcohol, the barbiturates and methaqualone. Since the latter have all been widely abused, it can be assumed that the effects they produce are not entirely unpleasant. Whether they have aphrodisiac effects in the strict sense is another matter. A drowsy and relaxed female is more likely to be receptive to the sexual advances of a male, but the ensuing performance is likely to be diminished rather than enhanced by a drug which dries up the body's natural secretions and eventually produces anaesthesia.

There is a general tendency for drug takers to accentuate the positive effects of the drugs with which they experiment, and methaqualone, as the heir apparent to mandragora, has undoubtedly gained a considerable reputation as an aphrodisiac (see below). In earlier times the use of an aphrodisiac was not necessarily aimed at improving the physical pleasure for an already compliant partner, but more often to help render an unwilling partner more susceptible. This practice is still common today, but with ethanol replacing the somewhat riskier and more toxic mandragora. There have been a few reports of drug users attempting to smoke datura leaves as a substitute for cannabis. The results have usually resulted in temporary hospitalization and an unpleasant period of detoxification.

Methaqualone

This drug strictly belongs in the section covering drugs of abuse, but since it shares some of the properties of mandragora as well as its root name, it will be considered here. Perhaps there was no ulterior motive involved when, following the introduction of methaqualone in 1965 by the William H. Rorer Co., one of the alternative trade names chosen for the product was 'Mandrax'. Probably it was only intended as a subtle acknowledgement of the historical use of mandragora as a 'drowsy syrup' or sleep-inducer. In the event, the manufacturers could hardly have anticipated the response that occurred among the young drug users of North America.

Mandrax itself was a mixture of methaqualone and diphenhydramine (an antihistamine with sedative properties). In America it appeared under the names of Quaalude, Parest, Sopor, etc. in a wide range of preparations. It was presented to the medical profession as a non-barbiturate, non-addicting sedative–hypnotic with no initial excitation phase and no morning-after hangover effects. From a pharmacological point of view it did have some advantages over the barbiturates which were, at the time, the principal drugs used in the treatment of sleep disorders. It did not produce a 'natural' sleep, but neither did the barbiturates, and the latter were beginning to cause problems from their tendency to produce tolerance (that is, increasing doses were required to achieve the same level of effect) and dependence.

The one feature of methaqualone was that it produced a distinct period of euphoria before sleep developed about 30 minutes after taking the tablets. Any drug that produces euphoria has a great abuse potential, and those in the know soon appreciated that if one concentrated on keeping awake, the euphoric phase could be prolonged. This became known as 'luding out'.

Illicit drug users are notoriously fashion conscious in their taste for new sensations, and the appearance of Quaalude or Mandrax in the home medicine cabinet soon created an epidemic of abuse. This was certainly helped by the extravagant newspaper headlines of the time: 'The love drug, heroin for lovers' said the *Washington Post* of 12 November 1972; 'The Dr Jekyll and Mr Hyde drug, hottest drug on the streets', claimed the *San Francisco Chronicle* on the following day. With free publicity of this kind it is hardly surprising that methaqualone-containing preparations soon presented a major problem of abuse.

From its neglect in the mid 1960s, methaqualone rose to be the most popular street drug by 1973, overtaking such previous favourites as cocaine, cannabis, LSD and heroin. As a result of several self-poisonings, methaqualone was added to the list of drugs in the UK included in the Drug, Prevention of Misuse Act. The motive behind taking any drugs of this type is to achieve the trance-like feeling of euphoria which they can induce. Methaqualone's action has been described by its devotees[8] as: '. . . a stumbling euphoria and a sense of not caring after you fight off the initial drowsiness', or, more precisely: 'Quaaludes totally intensify orgasm and allow me to concentrate harder on giving and receiving pleasure — makes me insatiable. I can freak all night.'

Less pleasant symptoms reported by regular users of the drug included headache, anorexia, nausea, abdominal cramps, and disturbed sleep patterns. In spite of these effects, methaqualone was especially popular among heroin addicts as a cheap and readily available alternative to narcotics, since it also could produce a 'high'. It was this 'high', together with the euphoria that was possible by mixing methaqualone with other central depressant drugs such as alcohol, that made it such a dangerous drug of dependence. The unusual feature of methaqualone is that, although regular dosing produces a tolerance to its sedative effects, the lethal dose still remains the same, about 125 mg per kilogram body weight. Since it is potentiated by alcohol, this combination represented an occasionally lethal mixture. In one treatment centre in the UK in the late 1960s, overdosage and toxic symptoms following methaqualone represented 15 per cent of all the cases admitted for treatment.[9]

Methaqualone is no longer available, even on prescription, and the benzodiazepine group of sedative–hypnotics have almost completely replaced all the earlier drugs used for this purpose. Although the newspaper headlines might have suggested the contrary, there is no firm evidence that methaqualone was an effective or reliable aphrodisiac. It seems more likely that it merely had a less detrimental effect on sexual performance than the other sedative–hypnotic drugs that were being misused at the time. Among young people in Europe, where alcohol happens to be relatively expensive, these types of drug can provide a cheap alternative to alcohol, although they are considerably more dangerous.

Alcohol

History of the Use of Alcohol

The great majority of cultural groups throughout the world have made use of alcoholic beverages of some sort or other. Many tribal peoples have used alcohol specifically in group rituals or religious ceremonies in order to gain spiritual experiences from the intoxication produced. Other 'mind expanding' or psychedelic drugs have also been employed for the same purpose, notably the sacred cactus *peyote* used by the Mescalero Apache Indians of North America, from which the drug mescaline has been extracted. Alcohol can in fact be considered to be the first and certainly most widespread drug used by man to alter his state of consciousness. The very ancient origins of the social consumption of alcohol can be inferred from the code of laws of Hammurabi of Babylonia about 1700 BC, in which the sale of wine was strictly regulated and riotous assemblies on the wine seller's premises were forbidden. These restrictions seem very similar to current licensing laws as practised today in England. There are also numerous references to the use of alcohol for both religious and secular purposes in the Bible.

'Until I come and take you away to a land like your own land, a land of corn and wine, a land of bread and vineyards . . .'

II Kings, 18.32

So wine was considered an important product and vineyards were clearly a valuable asset. However, the prophets also warned against over-indulgence:

'Woe unto them that rise up early in the morning, that they may follow strong drink; that continue until night, till wine inflame them!'

Isaiah, 5.11

The prophet Hosea is more specific in his warning:

'Whoredom and wine and new wine take away the heart.'

Hosea, 4.11

The association of drinking and loose sexual morals has apparently a long history. It is a feature of all societies where the custom of drinking has been adopted, and there is also an awareness of the dangers and social problems that result from inebriety.

In more recent times western civilization has spread the drinking habit throughout the world, and, with the exception of the strict adherents to the Muslim faith and some smaller temperance groups, alcohol is now the universal solvent for lubricating social intercourse.

There is no doubt that one important factor in the consumption of alcohol in a social context is its conscious or unconscious use as an aphrodisiac. Indeed, it is impossible to imagine western society without alcohol, and the protracted effort of the United States government to prohibit its sale between 1919 and 1933 proved disastrous. Alcohol has many of the features of the most pernicious garden weed: once introduced into a culture it is virtually impossible to remove, it is resistant to the most stringent measures of control, and it tends to spread rapidly. To take one example, the Scandinavian countries, and Finland in particular, where alcohol abuse is the greatest problem, are also those where alcohol is most expensive and restricted in availability. In contrast, in mediterranean countries where alcohol is extremely cheap and widespread, alcohol abuse and drunkenness have not, until recently, become major problems, although alcohol-related diseases such as cirrhosis of the liver do have a relatively high incidence. This lack of abuse may be due to the custom of drinking as an accompaniment to meals in a family environment rather than as a purely social activity or as a means of escaping from stress. However, research during the decade to 1980 has indicated that in Italy and Greece there is now an increasing incidence of alcoholism. Greece, which at one time had a less serious alcohol problem than almost any other European country, now has an estimated incidence of alcohol problems or alcoholism in 3 per cent of the adult male population and 0.4 per cent among females between the ages of 20 and 64.[17]

The Nature of Alcoholic Drinks

Since many of the aphrodisiacs available commercially contain alcohol as a solvent, and the effects of alcohol are very much determined by its concentration in a drink as well as its rate of consumption, it is worth considering briefly the nature of the drug.

Strictly speaking, alcohol is ethyl alcohol (ethan-2-ol, $CH_3 CH_2 OH$), a small organic molecule and only one of a series of alcohols of differing molecular structures. The vague use of the word 'alcohol' has occasionally led to tragic consequences since the closely related methyl alcohol (methanol, $CH_3 OH$) is extremely toxic and is metabolized in the body to form products which cause nerve damage, blindness and death. Illicit drinking of methyl alcohol, known also as wood alcohol, tends to occur in prisons or other institutions where ethyl alcohol is not readily available.

Ethyl alcohol is produced by a natural fermentation process by wild yeasts which, under the right conditions, convert sugars into alcohol and carbon dioxide. Wine is still produced by pressing whole grapes which have wild yeasts on their skins and these ferment the natural sugars released by the pressing. Beers, on the other hand, are produced from malted barley to which specific strains of yeast have been added. By these means it is possible to produce more or less palatable drinks containing between 2 and 18 per cent alcohol by weight. At higher alcohol concentrations the yeast is killed and fermentation ceases.

In order to obtain straonger drinks it is necessary to take advantage of the fact that alcohol boils at 70°C as against 100°C in the case of water. The art of boiling off the alcohol from a solution and condensing it to make a more concentrated alcohol solution (distillation) was invented by the Arabs and introduced into Europe by the returning crusaders in the twelfth and thirteenth centuries. Distillation enables one to obtain spirits containing anything from 18 to 95 per cent alcohol. (Pure alcohol tends to absorb water very readily and so special drying processes are needed to achieve 100 per cent alcohol.)

There are characteristic distilled alcoholic beverages or spirits produced throughout the world but principally in the temperate zones. The high esteem in which these drinks are held is sometimes reflected in their names. Whisky, for example, is derived from the Gaelic *usquebaugh* meaning 'water of life'. The same derivation can be seen in the Scandinavian drink *akvavit*. In tropical climates, less strong and more refreshing drinks are preferred, although some frighteningly strong spirits are produced in the West Indies from fermented and distilled sugar cane.

Most commercial spirits contain 37–40 per cent alcohol (70° proof spirit). Although stronger drinks can be made, they become less palatable as a result of the burning taste of the alcohol and are therefore usually diluted with other drinks. Alcohol itself does not contribute a significant taste to drinks − its burning effect can hardly be termed a taste − but it does bring out the flavour of other constituents or congeners in drinks. These congeners, which contribute to the taste and smell of drinks, are not always without physiological effects themselves. The wormwood originally used in absinthe, for example, was found to be a long-term neurological poison responsible for many of the toxic side-effects experienced by devotees of this particular aperitif. Consequently, the drink was banned and the non-toxic anise substituted in the traditional French *pastis*.

Why Do People Drink?

This is a question which has been given serious attention by both sociologists and psychologists as well as by clinicians for the last 40 years. It is still surprisingly difficult to answer. There are obviously a host of possible answers and it is a matter of assessing their relative importance. The following list has been culled from a number of cross-cultural sociological studies:

> to reduce anxiety;
> to overcome feelings of depression;
> to gain self-confidence;
> to escape from reality;
> to relax;
> to be sociable and not appear impolite;
> to get drunk;
> to forget;
> to appear sophisticated;
> to slake thirst;
> to obtain medicinal effect.

These reasons have not been listed in order of priority, but the anti-anxiety effect of alcohol appears to be the most important in a social context and is clearly related to self-confidence and relaxation in a potentially threatening environment. When these objectives have been achieved in the setting of a sexual relationship, then the potential aphrodisiac effect of alcohol is obvious.

Anxiety, fear, tension, depression and isolation are all factors that can induce or exacerbate a sexual problem. The alleviation of one or more of these states by moderate drinking might be thought by the individual to be clearly beneficial. Unfortunately, for a significant proportion of the population, social drinking leads gradually and inevitably to excessive heavy drinking and alcoholism. The effect of heavy drinking on sexual function is obviously not beneficial. In the short term it is overtly anaphrodisiac in the male and, in the long term, can lead to serious endocrinological or hormonal disorders.

Alcohol and Sexual Function

Until very recently, medical textbooks could add little useful information on the subject of alcohol and sexual activity except to quote the Porter from Macbeth:

Macduff: What three things does drink especially provoke?
Porter: Marry, sir, nose-painting, sleep and urine. Lechery, sir, it

provokes and unprovokes; it provokes the desire, but it
takes away the performance.

Macbeth II.iii

That almost four centuries of further scientific endeavour could add
virtually nothing to this description of the effects of alcohol is a tribute
to the acute observation of Shakespeare. Nose-painting refers to the
flush induced by drink as a result of the dilatation of the superficial
veins in the skin. The soporific and indeed anaesthetic properties of
alcohol are well documented and probably experienced at first hand at
some time by all regular drinkers. With regard to urine, the increased
ouput or diuresis is also a well-known nocturnal consequence of an
evening's drinking. The loss of body water over and above the volume
consumed is an important contributory factor to the dry mouth experienced the morning after drinking. Shakespeare's comment on lechery
aptly summarizes the paradoxical effects of alcohol which are, in
fact, dependent upon the amount consumed.

In seriously considering alcohol as a true aphrodisiac, it is important
to take into account the differences between the sexes in their response
to the drug, the dose of drug used, and the previous experiences of the
subject with alcohol. There is a critical dose at which alcohol is likely to
be effective, and this is determined by its rate of absorption and distribution to the brain. There is also a significant behavioural tolerance
to the effects of alcohol in subjects who are regular drinkers. For
example, a blood level of 300 mg/100 ml found in a drunken driver
stopped by the police would be sufficient to render a non-drinker
unconscious and incapable of even climbing into a car, much less
driving it. In trying to answer the question 'how can alcohol be an
aphrodisiac and an anaphrodisiac?', these factors must all be taken into
consideration.

The Pharmacology of Alcohol

The pharmacological properties of alcohol are principally sedative,
anxiolytic and anaesthetic. It also has analgesic properties and it is
quite common for inebriates to injure themselves while under the
influence of alcohol and yet be completely unaware of their injuries.
Another contributory factor to the apparent indestructibility of the
drunk is the muscle-relaxant effect of alcohol. Most of the damage done
in falling over occurs as a result of tensing the muscles before landing.
The uncoordinated drunk has a relaxed musculature and consequently
does less damage to himself when falling. Alcohol acts like any other

Figure 5.5: The Relationship Between the Amount of Alcohol
Consumed, the Blood Alcohol Concentration, and the Symptoms of
Intoxication. The shaded area represents the range that would include
approximately 94 per cent of the population. A unit of alcohol is
defined as being equivalent to the alcohol present in either: one half-
pint of beer (284 ml), one glass of wine (125 ml), or one measure of
spirits (one-sixth of a gill, 22 ml). The graph is intended only as a
guide: the actual peak blood level of alcohol achieved is very dependent
upon age, sex, weight, previous food intake, and the rate of
consumption of the alcohol, in addition to the number of units
consumed

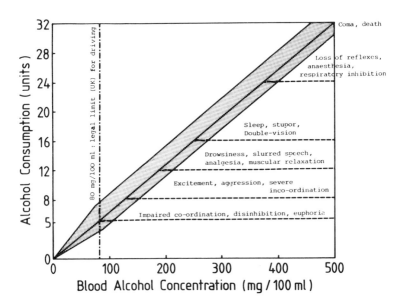

sedative–hypnotic or anaesthetic drug and there is a well-defined
spectrum of effects, which occur in a dose-dependent manner (see
Figure 5.5) as follows:

(1) *Excitement phase:* this occurs as alcohol blocks the inhibitory
nerves in the cerebral cortex of the brain (disinhibition), which results
in exaggerated and extroverted behaviour and loss of social inhibitions.
The normally intuitive moral controls become less evident and subjects
act in a more reflex and less controlled way. The ability to make
rational decisions is also diminished. Latent aggressive tendencies can

become unmasked and lead to overt violence. There have been several reports showing that a large proportion of sexual offenders are inebriated when they commit their offences.[10] In some cases this may only be an excuse given in court as an extenuating circumstance, but there is no reason why latent sexuality and aggression should not be simultaneously released by alcohol.

The value of alcohol as a social lubricant, helping shy people to talk and act more freely, is dependent upon this disinhibiting effect. Drunken behaviour is also highly infectious, and even non-drinking members of a party can appear uninhibited although they are completely sober. The great difficulty, however, is knowing when to stop drinking. The blood alcohol level, which determines the level of inebriation, will continue to rise for a time after stopping drinking as the alcohol already in the stomach continues to be absorbed.

The cocktail party is an extremely well-conceived device for rapidly bringing people to the excitement phase of inebriation since the drinks made available are usually those containing about 20 to 30 per cent alcohol (i.e. sherry, martini, cinzano, gin and tonic, etc.). This concentration of alcohol is the most rapidly absorbed from the stomach; more concentrated solutions (e.g. neat whisky) occupy a smaller volume and come into contact with a smaller area of the stomach lining, and less concentrated solutions (e.g. beer) have a much larger volume and, being more dilute, the rate of absorption of alcohol is slower. Cocktail parties are held at a time when people have empty stomachs so that the presence of food does not slow down the rate of absorption. In addition, such parties usually only last for an hour or two at most so that the guests do not generally consume sufficient alcohol or have enough time for their blood level to reach the next phase:

(2) *Motor-inco-ordination phase:* this phase is characterized by slurred speech, slowing of the reaction time and, eventually, loss of balance. (By this stage of a party most people are probably sitting or lying down and are unaware of the symptoms.) The hypnotic (sleep-inducing) effects of the alcohol become more evident and subjects feel drowsy and lethargic. If the volume of drink consumed is sufficient, it is at this stage that the emetic actions of alcohol may become apparent. Although an unpleasant side-effect, the act of being sick is not life-threatening unless the quantity of alcohol consumed is sufficient to produce the third phase:

(3) *Loss of protective reflexes:* the normal protective reflexes such as blinking and coughing are not under conscious control and are fairly resistant to the effects of sedative–hypnotic drugs. However, the combination of vomiting with loss of the cough reflex can result in the inhalation of vomit into the lungs with possibly fatal consequences from immediate asphyxiation or eventual pneumonia. These are a relatively common cause of death among heavy drinkers after excessive drinking bouts. The final stage of intoxication is:

(4) *Respiratory paralysis:* the breathing reflex, which is largely controlled by the rise in the concentration of carbon dioxide in the blood, is essential to life and can only be blocked by very high doses of alcohol or other hypnotic drugs. Most people fall unconscious (and therefore stop drinking) before a lethal dose has been absorbed. Nevertheless, there are occasional cases in which people have died from respiratory paralysis while in an alcohol-induced stupor.

In spite of the toxic effects of excessive alcohol just described, it is a relatively safe drug apart from its addictive properties, and its social usefulness lies in its lack of potency. Comparing alcohol with a sleeping drug such as Mogadon, for example, the blood concentration of alcohol required to produce significant sedation is about 20 000 times that of Mogadon. In other words, alcohol is 20 000 times less potent. The other factor which is important in considering the value of a sedative-hypnotic such as alcohol as an aphrodisiac is the safety margin between the dose that has a therapeutic effect and the dose that produces unwanted or toxic side-effects (e.g. loss of erection, impotence, headache, nausea, etc.). It is virtually impossible to take a sufficiently small dose of a sleeping drug of the barbiturate or benzodiazepine type to obtain only the initial excitement phase without sedation and sleep. Indeed, these drugs have been designed with the primary objective that they should produce a rapid and smooth induction of sleep. With alcohol the blood level at which subjects become aware of behavioural effects is about 50 mg/100 ml, although there is a wide variation between individuals. The legal limit for driving in the UK is 80 mg/100 ml and the level at which severe intoxication occurs is about 250 mg/100 ml or more. The lethal dose is thought to be about 500 mg/100 ml, although heavy drinkers and alcoholics have survived higher doses.

The time course of the rise and fall in blood alcohol level following a standard dose of 1.5 g per kilogram body weight is shown in Figure 5.6. In order to take advantage of the aphrodisiac properties of alcohol

Figure 5.6: The Time Course of the Rise and Fall in Blood Alcohol
Level, the Plasma Levels of Testosterone and Luteinizing Hormone
(LH), and the Subjective Self-assessment of Intoxication and Hangover
Following the Ingestion of 1.5 g of Ethanol per Kilogram Body Weight
in Healthy Volunteers. The data have been derived from Figures 1 and
2 in Ylikahri, R. *et al.* (1974) *J. Steroid Biochem.* **5**, p. 655

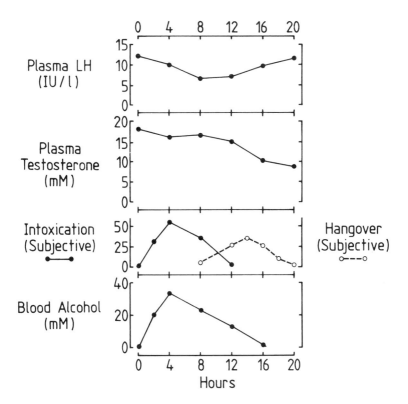

(that is, its unquestioned ability to provoke desire and release inhibitions), it is important to judge the dose exactly so that the detrimental
effects upon performance do not become apparent.

Sex Differences in Drinking and the Effects of Alcohol

It is an interesting paradox that it is only in recent times, in fact only
since the advent of the women's liberation movement in which sexual
equality is claimed, that scientific studies have examined sex differences
in the response to drugs, including alcohol, and found that significant

differences do exist.

The attitude of women to alcohol is determined to a great extent by their role in society. A survey in the United States in 1969 indicated that 77 per cent of the male population drank as compared with 60 per cent of the female population. Not surprisingly, the incidence of drink-related problems followed the same pattern. In earlier surveys the difference between the sexes was wider, so there is a tendency for women to catch up with men in this respect. Drink does not have quite the same 'masculine' image as, say, smoking used to have, and there is not much evidence that women in general drink in order to be accepted as social equals in male company. In adolescents, however, it is very clear that alcohol consumption is related to a wish to appear more adult or sophisticated. This tendency is probably common to both sexes. In Britain the men-only bar has virtually been legislated out of existence, and it is socially acceptable for women to drink in public. Nevertheless, the attitude to drunkenness in women is disapproving whereas the drunken male tends to be regarded with tolerant good will as an object of fun. Drunkenness in public is still considered as somehow unfeminine.

The effects of alcohol are also different between the sexes. Women are widely believed to be more susceptible to strong drink, but is this strictly true? Remembering that women on average weigh about 15 kg less than men, and that alcohol is distributed throughout the body, a fixed dose of alcohol will result in a higher blood alcohol level in women than in men. All other things being equal then, women are likely to be affected more by alcohol than men. Regular drinkers develop a behavioural tolerance to the effects of alcohol and, conversely, the completely inexperienced drinker is more likely to exhibit exaggerated signs of inebriation. As with all drugs that affect or alter mood, an expectation of the drug's effects plays a large part in determining the eventual response to the drug. This expectancy contributes to the placebo effect and can be strong enough for a subject to experience intoxication with levels of alcohol that would normally be without effect. In a study carried out in Bristol in which medical student volunteers were given standard doses of alcohol (corrected for body weight) or placebo drinks, the females appeared more inebriated than the males and became more garrulous and excited. At the same time the alcohol had less effect on their visual acuity and motor coordination.[11] Expectation and anticipation clearly played a large part in the females' psychological response to the alcohol, but they were unable to simulate the physiological effects, which involve less conscious control.

In considering the effects of alcohol on sexual arousal and function, the psychological and physiological effects of alcohol must be differentiated. The physiological response will be determined by the rate of drinking, the total amount of alcohol consumed, and whether or not food is present in the stomach. The psychological response will depend also on the nature of the drink and the environment. The peculiar efficacy claimed of champagne, for example, depends upon its reputation as an expensive and luxurious drink. The sparkling property of the champagne is also important since the bubbles of carbon dioxide act as a stimulant to the palate as well as being visually exciting. Champagne, in moderate doses, does not therefore have the same sedative actions of many other drinks, and can be positively stimulating. Of all the forms of alcohol, champagne is probably the most likely to be successful as an aphrodisiac.

Investigating the Effects of Alcohol on Sexual Function

There are basically two types of human study that can be carried out and neither one is ideally suited to the investigation of the effects of alcohol on sexual function. The first type is a self-report study. This involves choosing a group of subjects (males, females, drinkers, non-drinkers) to report on their sexual activity by filling in questionnaires. Although this enables sexual activity to be assessed under natural, as opposed to laboratory conditions, it has several limitations.

First, it is almost impossible to match subjects for social background, education and standard of living, all of which are factors that are likely to influence their attitudes to drinking and to sex. Secondly, it is not possible to regulate or control the level and frequency of drinking so that the 'drinker' group may include the regular light social drinker (10 to 12 pints of beer per week, for example) and the occasional binge drinkers who consume the same amount on average but in a concentrated session at weekends. Asking subjects to report their average alcohol intake does not therefore take into account the pattern of their drinking although this could very much influence their sex life. Finally, it is impossible to ensure that subjects report completely and truthfully, particularly with respect to their sexual activity, which is a subject that is embarrassing to most people. The separate interrogation of married couples or sex partners may reveal discrepancies, but it is still not possible to rule out collusion. Most studies of this type take a fairly long time to complete, and the circumstances of the subjects may change over the duration of the study. Masters and

Johnson[12] produced the classic report on human sexuality in a massive study of this type but unfortunately they were not investigating the influence of alcohol and so their reports contain little of immediate relevance to the subject of aphrodisiacs.

The second type of study is that carried out under laboratory conditions, which hopefully eliminates some of the variables in self-report studies. However, for both ethical and practical reasons, it is not possible to examine sexual activity in quite the same way or to determine the long-term effects of chronic drinking. Nevertheless, several ingenious physiological devices have been developed which enable sexual arousal to be measured in both sexes. Since an increase in physical arousal or erotic feelings is an important means by which aphrodisiac effects may be manifested, these studies have provided some useful information on the effects of alcohol.

Results from Self-report Studies

From the limitations outlined above, it is not surprising that contradictory or negative results have often been obtained in the few investigations that have been published. However, in a fairly large study (over 400 subjects) carried out by Beckman,[13] some interesting findings emerged. The study reported the percentage of alcoholics and non-alcoholics of both sexes desiring, having and enjoying sexual intercourse when drinking or not drinking. Although the study was aimed principally at investigating alcoholism, it did provide some useful information because each subject was acting as their own control. That is, the non-alcoholics reported under non-drinking, slight drinking and heavy drinking circumstances. A particular difficulty, mentioned by Beckman, was that a sober woman may report that she believes that her sexual enjoyment is increased after drinking, although she may not report an actual increase in enjoyment under the same conditions. This contradiction well illustrates the preconceptions held concerning the effects of alcohol which, in many cases, differ quite widely from the actual effects achieved.

Beckman's sample consisted of 477 subjects aged between 20 and 59 years of whom 120 were female alcoholics and 119 female non-alcoholics. The details of their sexual activity revealed that female alcoholics were more likely to have sex relations with different people while drinking and were more likely to indulge in a variety of sexual acts. The male alcoholics showed similar differences compared to non-alcoholics but, in addition, indicated that they were more likely to have intercourse with partners they would not have become involved with if they had not been drinking. The males (both alcoholic and non-alcoholic) reported very little desire for, or participation in, sexual intercourse when drinking a lot. There were no differences in the frequency of desire, participation, or enjoyment of sexual intercourse between non-drinking and light drinking conditions.

These observations suggest that in the male, a little alcohol has no obvious effect, but that heavy drinking has a significantly detrimental effect on the libido. Returning to the female groups, the results suggest that alcoholics are perhaps more promiscuous than normals and are also more likely to believe that increases in sexual enjoyment, desire and frequency occur when drinking. However, about 40 per cent of the women questioned did not share this view, and the general conclusion must be that alcohol alters one's perceptions of sexual ability and pleasure. Women with pre-existing sexual problems are perhaps more likely to turn to drink as a means of relieving them and, in this respect, a spurious correlation between drinking and increased sexual gratification can be derived.

Laboratory Studies on Sexual Responding

The physiological methods available for measuring sexual arousal in either sex are mentioned in Chapter 6. Most studies with alcohol have involved only male subjects, but some of these have been well designed and have produced useful information. Although the report by Rubin and Henson[14] involved only 16 male subjects aged between 20 and 29 years, they were all healthy volunteers with no previous sexual or alcohol-related personality disorders and so might be considered to be more representative of the population as a whole.

The design of the experiment was a double-blind crossover (see page 169) and a serious effort was made to disguise the presence of alcohol in the drinks, which ranged from concentrations equivalent to a dose of 0.5 ml/kg (about 1½ pints (750 ml) of beer) to 1.5 ml/kg. The criterion for sexual arousal was defined as a penile erection and was measured using a strain gauge transducer which recorded the circumference of the penis. Baseline flaccidity was designated arbitrarily as zero and a full erection as 100 so that, at any time, the degree of penile erection could be expressed as the percentage of a full erection. The erotic stimuli to which the subjects were exposed consisted of 10-minute black-and-white video films illustrating a variety of heterosexual acts (the type of films that we would think of as 'blue' or 'stag' films). Each subject watched four different films each twice, first under the instruction to relax and allow an erection to develop if possible, and then to mentally inhibit his erotic feelings. Alcoholic or placebo drinks were consumed after the second presentation of each of the first three films. After the last presentation (film 4), the subject was requested to try to become sexually excited without any external stimulus or physical assistance. This was termed 'phantasy eroticism'. As an additional check to make sure that the subjects actually watched the films, they had to give a signal response to occasional flashes of light projected on to the viewing screen.

The three parameters recorded were (1) peak erection, (2) average erection, and (3) time taken for the attainment of a 20 per cent erection. The main limitation in this particular study was that the blood alcohol levels were not monitored. Although the subjects received standard doses of alcohol, it would have been very interesting to know whether or not the effects of the drink on sexual function corresponded in time to the peak blood alcohol concentration. The results indicated that, following the low or moderate doses of alcohol (0.5 or 1.0 ml per kilogram body weight), a small yet significant reduction in average sexual arousal occurred but that the conscious suppression of arousal or 'phantasy arousal' was unaffected. After the high dose, however, all three parameters were inhibited, although none of the alcohol doses affected the subjects' ability to supress sexual responding.

This study only measured one, physical, aspect of sexual responding in one sex. However, the penile erection is the most important factor from the male point of view in achieving a successful sexual encounter. Its suppression by alcohol at fairly moderate doses confirms the long-held belief that alcohol can take away the performance.

Alcohol as a Local Anaesthetic

It is worth considering the importance of the local anaesthetic effects of alcohol since, under normal circumstances, the ejaculatory or orgasmic phase of sexual arousal is encouraged by mechanical stimulation of the penis by one means or another. In situations where premature ejaculation makes sexual intercourse less enjoyable or even impossible as a result of the post-ejaculatory loss of erection, a delayed ejaculation due to alcohol could be beneficial. Any effect of alcohol on the time to ejaculation would be very difficult to assess accurately even with the use of more sophisticated technology that might make it possible to provide a standardized mechanical stimulus to the penis. The ejaculation reflex, although once initiated is not under conscious control, can easily be suppressed or completely inhibited by feelings of anxiety, fear or lack of concentration. Again, the anti-anxiety actions of alcohol could be helpful here also. The principal risk lies in consuming too much alcohol so that an erection becomes impossible in the first place.

From the female point of view, premature ejaculation is also highly unsatisfactory. Delayed or inhibited orgasm as a result of alcohol is not such a problem, but a reduction in vaginal sensation could be helpful in cases where intercourse has previously proved painful or uncomfortable and the discomfort has inhibited the sexual drive. For women, the relaxing and disinhibiting influence of alcohol would be an aid to successful intercourse, and the risk of unintentionally exceeding the optimum dose not as potentially catastrophic as in the male. On balance, therefore, it appears that the Porter's advice to Macduff was accurate and pertinent to that gentleman, but his advice to the opposite sex might well have been different.

Alcohol as an Anaphrodisiac

Having considered the potentially aphrodisiac properties of the universal solvent, it is only right to examine the other side of the coin and consider its toxic properties. There are several means by which alcohol can act to reduce the libido and the ability to indulge successfully in sexual activity.

In the short term, excess alcohol acts as a general depressant throughout the body and nervous system. Since the achievement and maintenance of an erection in the male is dependent upon a fully functional autonomic nervous system controlling the blood flow to the penis (see page 147), it is not difficult to imagine how alcohol might act to take away the performance. Although similar blood-flow changes occur in the female sex organs during arousal, they are not so critical for the completion of successful intercourse. In the female, the anaesthetic effects of alcohol may reduce sensation to the point at which an orgasm is impossible. In this respect alcohol can be said to act to reduce sexual satisfaction and cause frustration. As mentioned before, from a male point of view, the object would be to delay orgasm without completely inhibiting it.

The hangover effects of alcohol are well known, and the 'morning after' syndrome is the recognized risk following a heavy night's drinking. The symptoms include dehydration and dry mouth (produced as a direct result of the diuretic effect of alcohol), headache, throbbing temples, nausea, muscle tremor and hypersensitivity to light, sounds and smell. The relationship between the time course of intoxication, hangover effects and the blood alcohol concentration has been investigated in a study of healthy volunteers by Ylikahri *et al.* in 1974.[15]

The subjects reported their subjective impression of their degree of intoxication and the severity of the subsequent hangover using the 10 cm linear scale method described below. The results are shown in Figure 5.6. The dose of alcohol used (1.8 ml per kilogram body weight) was relatively high for this type of study and was obviously quite sufficient to produce a noticeable hangover in most of the subjects.

The assessment of intoxication matched exactly the blood alcohol concentration both in time course and peak effect. The hangover effects, however, did not. There also appeared to be some correlation between the individual subject's sensitivity to the immediate effects of the alcohol and the severity of the hangover that followed. There is still some doubt as to the extent to which alcohol itself accounts for the unpleasant after-effects. The major metabolite produced in the liver from alcohol is acetaldehyde. This is an extremely toxic substance, but fortunately it is, in turn, rapidly metabolized to acetate, which is non-toxic and can be used as fuel by the body's cells. (This is the biochemical basis for the rather ambiguous claim that alcohol is 'food'. It does, however, contain 7.7 calories per gram.)

Experiments with very low doses of acetaldehyde in volunteers showed that it could mimic most of the symptoms of the hangover but without the inebriating effects of alcohol. However, the ability of different drinks to produce hangovers varies widely and there is also the long-standing belief that mixing drinks tends to produce worse hangovers. The reason for these differences is thought to be the presence of other substances or congeners in the drink, which can also contribute to the hangover. Evidence for this belief comes from the observation that the 'purest' drinks (that is, those having the lowest concentration of congeners, such as vodka, gin and white wine) have the lowest hangover potential. On the other hand, the more complex mixtures with high levels of congeners (brandy, whisky, red wine and sherry) tend to produce more hangover symptoms for a given amount of alcohol. There are individuals who might, from personal experience, dispute these categories, but some people do show idiosyncratic reactions to specific drinks such as gin, and for them it is best to find out the safest drink by a process of trial and error. There is, incidentally, no scientific basis for the belief that mixing drinks increases the hangover potential; the physical state of the drinker on the morning after is determined only by the amount of alcohol consumed and the nature of the drinks.

Chronic Drinking: Sexual Problems Associated with Alcoholism

The popular view has it that chronic male alcoholics are cheery bucolic individuals or sad dropouts having little or no interest in the opposite sex, whereas the female alcoholic is more likely to be of loose morals. Kinsey,[16] writing in 1966, cited the clinical observation that alcoholics are devoid of sexual tension. This presumably means that they are not worried about it, perhaps because their principal concern is to determine where the next drink is coming from. The lack of interest in sex may therefore be a state of mind rather than a symptom of physical deterioration. Alcohol dependence, although cheaper to maintain than a dependence on hard drugs, does require considerable financing, and some female alcoholics turn to occasional prostitution as a means of raising funds for drink.

It is also possible that women who are promiscuous for reasons of anxiety, loneliness or self-doubt are quite likely to turn to drink as a means of relieving their symptoms. Kinsey, in a study of 46 hospitalized alcoholic women,[16] found that 33 of them reported frigidity in sexual relations. One of the reasons given to the investigators by women who have become alcoholic is that they began drinking as a result of an unhappy marriage. In women it would therefore appear that alcoholism occurs as a consequence of unsatisfactory sexual relationships rather than as a root cause of sexual problems.

The hormonal changes that accompany chronic alcohol intake have been studied far more thoroughly in men than in women and it is well established that alcoholism is associated with atrophy of the testicles and reduced fertility. Since testosterone, the male sex hormone primarily responsible for maintaining the libido, is produced by the testes, the testicular atrophy can lead to reduced circulating levels of the hormone and consequently to a reduced libido. What is of more concern to the healthy male drinker in the short term is the effect of acute doses of alcohol on sexual function. A recent study has shown that even a single dose of 1.5 g/kg of alcohol (roughly equivalent to 6 pints (3.4 litres) of beer) can produce a significant reduction in the circulating testosterone levels between 10 and 20 hours after drinking.[15] Although the libido and overall sexual function are not completely dependent upon testosterone levels, alcohol clearly produces hormonal disturbances that can affect in a deleterious way the normal functioning of the reproductive organs.

Chronic alcoholism is also associated with cirrhosis of the liver, gastritis (inflammation of the stomach lining), cardiomyopathy (damaged heart muscle), and diabetes. In pregnant women who drink heavily, the foetus can be damaged and suffer from what is known as the foetal alcohol syndrome. This causes characteristic physical deformities in the child and a later retardation of the physical and mental development of the child. It is an interesting thought that if alcohol were to be newly discovered by the pharmaceutical industry and given controlled clinical trials as a sedative or tranquillizer, it would not pass the stringent safety regulations and would be banned from the market.

Summary and Conclusions

Alcohol is undoubtedly the most widely used drug in western society and its pharmacological and toxicological properties have been extensively investigated. Having stated at the beginning of this chapter that alcohol is a relatively safe drug, it should now be clear that no drug is completely safe and that excessive consumption of alcohol, either acutely or over a long period, will produce unpleasant side-effects and possibly eventual dependence or addiction to the drug. It has been estimated that at least 1.5 per cent of the adult population of the UK is dependent upon alcohol; that is, about 850 000 alcoholics or problem drinkers, compared with the 3311 narcotic drug addicts notified to the Home Office in the same year (1981). This last figure is a gross underestimate of the real problem of drug abuse. Over the last few years customs seizures of illicit drugs have been increasing by between 75 and 90 per cent per annum.

Alcohol is drunk mainly to counteract anxiety and reduce tension and as a social custom. Under the appropriate circumstances it has definable aphrodisiac properties: it reduces inhibitions in both sexes and can therefore increase overt sexual desire. In the male it can delay orgasm, and in the female it can produce relaxation and possibly facilitate intercourse by its anaesthetic and muscle-relaxant properties. There is no evidence that alcohol can increase sexual satisfaction directly in either sex, and at increasing doses it becomes distinctly anaphrodisiac in the male. It is a unique drug and it is very unlikely that any other substance will be found with the same spectrum of actions.

6 THE SCIENTIFIC APPROACH TO SEX AND APHRODISIACS

Endocrinology of Sexual Function

It has been known since antiquity that castration causes a loss of male sexual characteristics and a decline in the sexual appetite. The Chinese were aware 4000 years ago that giving large doses of testicular extracts could overcome some of this loss. The Pen Tsao herbal also recommended the use of semen from young men for the treatment of sexual weakness, although it was believed that the brain was the ultimate source of the sperm. The great Greek physicians wrote knowledgeably about conditions such as mumps which could adversely effect the functioning of the male organs. Aristotle wrote that the semen was the formative or activating agent, whereas the female element was merely passive and required fertilization by the sperm, and it was believed that the right testicle yielded male offspring and the left female. Animal testicles, in dried or powdered form, were recommended as aphrodisiacs by several authorities in Chinese, Greek and Arabic medicine.

The presence of spermatozoa in human semen was first identified in 1674 by Leeuwenhoek using his newly invented microscope. Until then it had been believed that the male seed consisted of a miniature man or homunculus, anatomically complete in every respect. However, the presence of male sex hormones in testicular extracts and their chemical identification were not achieved until the 1930s.

The ovaries were also known in ancient medicine and the Jews had long recognized that removal of the ovaries of cows would cause fattening and prevent breeding. Aristotle also described the removal of the ovaries in sows and camels for the same purpose. However, it was only in the seventeenth century that Stensen suggested the name 'ovaries' for these organs; up to that time they had been described as female testicles. The human ovum or egg was discovered at the beginning of the nineteenth century and, shortly afterwards, the union of sperm and ovum was observed. As with the male hormones, it was not until the 1930s that the female hormones were isolated, initially from urine and later from placental tissue.

In the 1820s Gall and Spurzheim, the founders of the so-called science of phrenology, advanced the theory that sexual impulses arose in the cerebellum of the brain. The principal scientific evidence for this bizarre idea consisted of some fairly subjective observations on a singularly amorous widow of Gall's acquaintance. Shortly afterwards, Berthold made the spectacular discovery that the transplantation of the testes of a cock into a castrated rooster would restore its comb back to its former splendour. This was the first practical demonstration of the efficacy of testicular tissue in restoring male sexual characteristics, and was, for many years, the best available assay for preparations of male sex hormones.

However, endocrinology as a science only really began at the turn of the century and was due, in large part, to the work of E.H. Starling, who coined the word hormone from the Greek *hormon*, meaning 'to arouse or excite'. The word endocrine is used to describe the activity of the glands in the body whose secretions are internal, that is, directly into the bloodstream. The production of semen is described as exocrine since it is secreted externally.

The Nature of the Sex Hormones

The sex hormones which are normally found in the body are all based on the same chemical structure, the steroid nucleus:

Before entering into a discussion of the possible role of the sex hormones in altering sexual function and the libido, it is probably worth describing their specific physiological properties.

Oestrogens

These are largely responsible for the development and maintenance of the female sexual organs and secondary sexual characteristics. They cause the fat deposition on the hips and breasts and the restriction of the spread of body hair in the female. They stimulate peripheral blood flow and can cause a water retention which becomes particularly evident in the weight gain that occurs immediately before periods, when the circulating oestrogen levels are increased. Their effects are modified by progesterone.

Progesterone

The interactions between this progestogen and the oestrogens are complicated; at times they can supplement each other, but at some sites in the body they act antagonistically. Progesterone is responsible for the establishment and maintenance of pregnancy. The use of progesterone in oral contraceptives is basically to make the body act as if it were pregnant and thus suppress ovulation and menstruation.

Testosterone

This is the principal androgen or male hormone synthesized by the Leydig cells in the testis and is responsible for the major body changes that occur at puberty in the male. The circulating level of testosterone in the blood is very low until puberty, when it climbs rapidly to a peak which is maintained over a period of between 25 and 45 years. After this time it gradually declines so that old age is associated with a lessening of the libido.

Control of Sex Hormone Release

The synthesis and release of the sex hormones is under the control of two gonadotrophic hormones secreted by the anterior pituitary tissue situated at the base of the brain. The first, follicle stimulating hormone (FSH), controls the ovarian cycle in the female. This results in the development of follicles and, consequently, mature ova in the ovary. It also regulates the development of mature sperm in the testis of the male. The second, luteinizing hormone (LH), appears to control ovulation, that is, the release of the mature ovum from the ovary. It also controls seminiferous tubule growth and androgen secretion in the testis. A third human gonadotrophin, chorionic gonadotrophin (HCG), is secreted by the placenta and is probably only significant during pregnancy when it suppresses menstruation.

The release of gonadotrophins from the anterior pituitary is in turn controlled by releasing factors secreted by the hypothalamus, the region of the brain situated just above the pituitary stalk. A complex feedback control system exists to regulate the gonadotrophins and sex hormones (see Figure 6.1). For example, in pituitary dysfunction in the male, low levels of LH result in low testosterone levels and deficient secondary sexual characteristics and reduced libido. The administration of exogenous testosterone as replacement therapy, rather than having a beneficial effect, will act to suppress even the low level of secretion of LH by the pituitary. However, replacement therapy with LH and FSH

Figure 6.1: The Physiological Control of Sexual Activity. This figure is based on Figure 1 in 'Drugs, Transmitters and Hormones and Mating Behaviour' by A.G. Karczmar (1980) in *Mod. Probl. Pharmacopsychiat.* *15*, 1. The sympathetic and parasympathetic nerves are the two divisions of the peripheral autonomic nervous system, which is not normally under conscious control. The motor nerves stimulate the skeletal muscles which are under voluntary control. (A more detailed explanation of how drugs may interact with the chemical neurotransmitters of the central and peripheral nervous systems is provided in Appendix 2.) *Key:* MEDE., median eminence; ARC., arcuate nucleus; MB., mamillary body (these are all specific brain regions close to the hypothalamus); ACH, acetycholine; NA, noradrenaline; DA, dopamine; 5-HT, 5-hydroxytryptamine (serotonin)

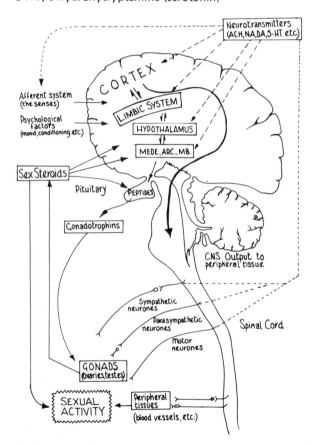

in the appropriate ratio will result in normal development of the testes and restored sperm production.

It is difficult to draw a precise line between what is the normal variation in endocrine function within the population, and endocrine disorders requiring treatment. The normal level of circulating testosterone in the adult male, for example, is given as a range of concentrations rather than as a precise figure. The treatment of endocrine disorders is usually either by replacement therapy if hormone levels are abnormally low or absent, or by the administration of hormone antagonists or suppressors of release in cases where the levels are too high.

A very large number of synthetic sex hormone derivatives have been made which are, in many cases, more potent or more long-lasting in their effects than are the parent compounds. They are used principally as oral contraceptives (oestrogens and progestogens), anabolic steroids (testosterone, dihydrotestosterone), or anti-androgens (cyproterone). They have a wide spectrum of activity and a host of clinical uses. The main therapeutic applications which are relevant to sexual behaviour have been summarized in Table 6.1.

Table 6.1: Clinical Applications of Steroid Hormones

Steroid	Dose	Therapeutic Object
Oestrogen	Low	Induction of ovulation in infertility. Treating symptoms of the menopause. Increasing cervical mucus.
	High	Treatment of hypersexuality in males.
Androgens	Very low	Treatment of female frigidity.
	Low	Treatment of constitutional impotence.
	Medium	Treatment of the male climacteric (equivalent to the menopause in women). Anabolic effects (increase of body weight due to nitrogen retention).
Progestogens	Medium	Oral contraception (the 'mini-pill'). Treating habitual abortion.
	High	For dysmenorrhoea, pre-menstrual tension and long-term contraception. Treatment of hypersexuality in males.
Oestrogen plus progestogen		Combined oral contraceptive. For postponing menstruation and for the treatment of acne.
Anti-androgens		Suppression of precocious puberty. Treatment of hypersexuality or deviant behaviour in males. Treatment of hirsutism in women.
Gonadotrophins (not steroids)		Induction of ovulation. Treatment of retarded puberty and eunuchoid syndromes.

Forms of Sexual Inadequacy

Many instances of depression, anxiety, insomnia and other disorders have as their root cause a fear or concern over some real, or even imagined, problem of sexual inadequacy. In fact, a low sexual drive or appetite only becomes a serious problem when the other partner in a marriage has a much greater appetite. A married couple with similar levels of expectation of sexual pleasure are very fortunate; unmarried couples are best advised to find an alternative partner with a more similar libido. A low sex drive may be an intrinsic part of an individual's personality, relying as it does on the activity of the higher centres in the brain, or it may occur quite naturally with increasing age. Throughout history, men have used aphrodisiacs in a vain attempt to reverse the natural tendency to a lower level of sexual activity in old age. The occasional exception to this rule, exemplified by the sprightly 70-year-old of average means who marries a girl of 20, still provokes whispered hints of monkey glands, rejuvenation treatments, or worse. The same gossips of three centuries ago would have muttered implications of witchcraft and other dark practices, particularly in cases where an older woman succeeded in attracting and keeping a younger man.

Apart from a low libido, the other principal cause for feeling sexually inadequate is impotence, the inability to produce an erection adequate for coitus. The Kinsey report indicated an incidence of impotence of 1.3 per cent in males up to 35 years old, increasing to 6.7 per cent at 50 and 18.4 per cent at 60. It is very rare to find a case of primary erectile impotence, that is, a case where an erection has never been achieved, but severe illnesses can produce temporary impotence, which disappears after a period of convalescence. A particularly traumatic event or illicit drug abuse can also lead to impotence of a more or less temporary nature. In this case the impotence is often symptomatic of a deeper psychological or pathological condition, and it does not respond to direct treatment. However, this does represent a situation in which self-medication by the indiscriminate use of counter-irritant, or rubefacient aphrodisiacs is common. Where there exists an underlying cause for the impotence, the attempted treatment of the impotence on its own can lead to more problems than it solves.

Secondary impotence, that is, impotence occurring after a period of normal sexual activity, can also be drug-induced. Masters and Johnson, in their textbook on human sexual inadequacy (see reference 12 in Chapter 5), provide a list of drugs which were implicated in the case histories of impotence reported up to that time. This list is being constantly updated as newly introduced drugs are found to have an-aphrodisiac side-effects. Although this topic is discussed in more detail later, it is perhaps worth listing here the groups of drugs known for their detrimental effects on sexual performance. They include: narcotic analgesics, alcohol, amphetamines, atropine, benzodiazepines, digitalis, guanethidine, mono-amine oxidase inhibitors, methyl-DOPA, pheno-thiazines, reserpine and nicotine.

Premature ejaculation is certainly the commonest sexual problem in the male. Masters and Johnson defined it as: 'the inability to control ejaculation for a sufficient length of time during vaginal containment to satisfy the partner on at least 50% of their coital connections'. It is a problem that occurs most frequently in early post-pubertal years, and usually becomes less of a problem with increasing practice and exper-ience. The opposite situation, that of ejaculatory incompetence, in which ejaculation is delayed or does not occur at all, is less common. It can happen after taking amphetamines or cocaine, and is one means of prolonging sexual intercourse. This can be an advantage to the woman, who might otherwise be left unsatisfied, but it is often frustrating for the male. It is not unusual to achieve an orgasm without an ejaculation, or vice versa, but many modern aphrodisiacs are sold as aids for the prevention of premature ejaculation and contain a mild local anaesthe-tic for this purpose. A rubber sheath would have a similar ability to reduce the sensitivity of the penis to tactile stimulation.

The Origin of the Erection

There are two basic ways in which an erection can be achieved. First, a reflex erection can be evoked solely by tactile stimulation of the eroto-genic zones. This is purely a spinal reflex similar to the knee-jerk reflex and does not involve the brain. Secondly, a psychic erection can be achieved as a result of the sensory inputs to the brain: for example, viewing erotic pictures or, more rarely, reading erotic prose. The reflex erection can be suppressed by inhibitory influences from the brain, but can still occur against the will. In cases of spinal injury, where there is no connection between the brain and the lower spinal cord, the reflexive erection is no longer subject to any inhibitory control from the brain, and the threshold for excitation becomes much reduced.

The psychic erection can be induced by sensory conditioning: some-times a hint of a perfume, or the texture of a material, is sufficient to provide an emotive recall of past sexual activity in a manner powerful enough to induce a physical response in the present. If this conditioning is carried to the extreme, in which the object itself replaces the normal sexual partner, the end result is a sexual fetish which can develop to the point when normal sexual relations become impossible without the object of the fetish. From a brief reading of the magazines that provide a forum for the discussion of sexual problems, there seems to be no end to the variety of substances or objects with fetish potential. The most well known are rubber and leather, but some very bizarre objects, including a stuffed parrot, have been recorded as sexual fetishes. In the widest sense it is possible to consider drug abuse associated with sexual activities as a form of chemical or pharmacological fetishism.

Physiology of the Erection and Ejaculation

The autonomic nervous system controls the automatic functions of the body by exerting a balance between the normally opposing sympathetic (adrenergic) and parasympathetic (cholinergic) systems (see Appendix 2). Erection occurs in response to the activity of the parasympathetic *nervi erigentes*, which arises from the sacral region of the spinal cord and causes relaxation of the muscular wall of the arterioles (small arteries) of the penis. As a result, the blood flow into the spongy tissues of the organ is increased, causing it to swell, and at the same time the veins which normally drain the blood from the penis are physically compressed, so trapping the blood within the engorged organ and enabling an erection to be maintained. There is some evidence from patients suffering from spinal injuries that the sacral centre is not essential for achieving an erection, and that other, sympathetic, nerves from the thoracic and lumbar regions of the spinal cord can generate an erection in the absence of the sacral region.

Ejaculation occurs in three stages, and although it is normally initiated by strong tactile stimulation of the erect penis, an erection is not essential in order for ejaculation to occur. Indeed, this can be one of the more distressing forms of male sexual inadequacy (see below). In the first stage, sympathetic nerves produce contraction of the smooth muscles of the internal genital organs sending semen along the *vasa deferentia* and into the region of the urethra between the internal and external sphincters of the bladder (see Figure 6.2). This is termed emission. Next, the internal sphincter of the bladder is contracted so as to prevent the retrograde passage of semen back into the bladder or the simultaneous discharge of urine. Finally, rhythmic contractions of

Figure 6.2: The Nervous System Control of Erection and Ejaculation. Details of the chemical neurotransmitters released by the different nerves are provided in Appendix 2

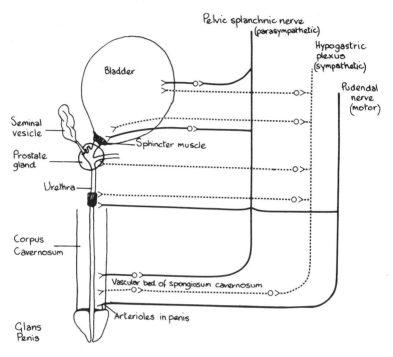

the bulbo-cavernosus muscle propel the semen, together with the accumulated secretions of the various accessory glands, along the length of the urethra to achieve ejaculation proper.

The act of ejaculation thus requires the complete functioning of both sympathetic and parasympathetic nerves as well as the motor nerves which pass directly from the spinal cord to the striated muscles of the penis. Physical injury, drugs, increasing age, and disease states such as diabetic neuropathy, can all have greater or lesser detrimental effects on the functioning of the nervous system and thereby interfere with sexual performance. Some of these factors will be considered in more detail in the next chapter, but male sexual inadequacy does not necessarily have a purely physical basis.

Male Sexual Inadequacy

Most men have an inflated view of their own sexual potency, and the exaggerated tales of sexual conquests which tend to be exchanged in male company often serve only to make the situation worse for the individual who is made to feel that his own level of sexual activity is somehow below par. If the truth were known, and the braggadocio divorced from reality, a more reasonable level of sexual expectation might be considered adequate by general standards. The famous Kinsey survey of 1948 served to illustrate this point: in males, the average frequency of all sexual outlets in all subjects was 2.74 times per week. In males aged 30 or under, no less than 2.9 per cent reported a frequency of orgasm of once in ten weeks or less. It is clearly wrong to suggest to someone who is perfectly satisfied by infrequent sex that he is in some way deficient, and that he should expect and achieve a higher rate of performance. It is the anxiety associated with such feelings of sexual inadequacy, however unjustified, that leads to a vicious circle in which a satisfactory sex life becomes increasingly difficult to achieve.

Current social fashions have generated an increasing pressure on both sexes to live up to an idealized level of sexual performance. The greater sexual awareness of the population, and the prevalence of adult magazines and journals that purport to provide some sort of sexual counselling, have led many people to believe that they are somehow inadequate. Some indication of the extent of this problem within marriage can be gained from the results of a survey of 100 'normal' couples published in 1978.[13] The couples, who were predominantly white, well educated and happily married, responded to a self-report questionnaire. Some of the results have been summarized in Table 6.2, and it is immediately apparent that sexual 'difficulties', as opposed to serious problems, are more frequent among normal married couples than one might expect.

The time is probably long past when a married woman, through a combination of ignorance and shyness, could go through married life without experiencing an orgasm, or even being unaware that such a natural phenomenon existed. In fact, the pendulum has swung the other way so that an increasing number of people are becoming sexually more adventurous and demanding of their partners. It is as though everyone wanted to be able to run a mile or win a marathon. Of course, with a suitable training programme, it is possible to increase one's physical performance, and there is no reason why sexual activity should not be considered as being like any other form of physical activity. Some less than serious studies have estimated that a short

Table 6.2: Frequency of Self-reported Sexual Problems (Derived from Frank, E. *et al.* (1978) *New Engl. J. Med.* **299**, p. 111)

Sexual Dysfunction			
Women	(%)	Men	(%)
Difficulty getting excited	48	Difficulty getting an erection	7
Difficulty maintaining excitement	33	Difficulty maintaining an erection	9
Reaching orgasm too quickly	11	Ejaculating too quickly	36
Difficulty in reaching orgasm	46	Difficulty in ejaculating	4
Inability to have an orgasm	15	Inability to ejaculate	0
Comparison of sexual difficulties reported by men and women:			
Partner chooses inconvenient time	31		16
Inability to relax	47		12
Attraction to person other than mate	14		21
Disinterest	35		15
Different sexual practices	10		12
'Turned off'	28		10
Attracted to person of same sex	1		0
Too little foreplay before sex	38		21
Too little 'tenderness' after sex	25		17

session of active sex is equivalent, in terms of the energy expended, to a half-mile jog. The use of various stimulant drugs or hormones to improve physical performance is not unknown in professional sport; these same drugs might be expected to have similar effects on sexual performance, a point which will be considered later.

Some people possessing a very strong imagination are able to achieve a psychic erection purely by thinking erotic thoughts and in the absence of any external sensory input. It is thought that it is the limbic areas in the brain which are responsible for mediating this type of erection. Some patients who have lost their penis, either through disease or accidental injury, still report being able to experience the feelings of an erection, and even an eventual orgasm, in spite of the missing organ. The brain clearly maintains a strong memory of the physical sensations of sexual arousal and can recall them at will.

Psychological influences on the ability to achieve an erection and orgasm are extremely complicated. In some individuals a certain level of anxiety is necessary if they are to achieve an erection; in others the slightest anxiety or embarrassment can effectively prevent it. Difficulties in achieving an orgasm can be directly related to a fear of causing an unwanted pregnancy, and even strict moral codes or religious orthodoxy have been implicated in some cases of sexual problems.

It is therefore not easy to make any general assumptions about the basic cause of impotence in any individual.

Treatment of Male Sexual Inadequacy

Methods of treatment for the various conditions already described can consist of drugs, hormones, psychiatric counselling, full psychoanalysis, or any combination of these. Some of the alternative medical practitioners also profer treatment for sexual disorders, and hypnotism, herbalism, and all shades of meditation are available for those patients who have either found traditional medicine to be ineffective, or are too shy to divulge the details of their intimate problems to their family doctor.

A doctor or clinician, in prescribing any particular drug or hormone to a patient, would probably not consider that he was employing an aphrodisiac *per se*, in spite of the fact that it is often a frank aphrodisiac effect that is being sought. There is still a general sense of embarrassment about 'personal' problems, which inhibits patients from talking frankly to a doctor, and which drives them to consult less reputable, but also less compromising sources for a remedy to their problem. In the treatment of both male and female sexual inadequacy, it is essential for the physician or counsellor to reassure and relax the patient so that any barrier to open discussion is overcome. In this way it should be possible to discuss the problem and seek the underlying cause without recourse to drugs.

A large number of jokes still derive from the situation of the worried man seeking an interview with his doctor to discuss some private problem, and being given some anonymous pills with the advice that they will 'see him all right'. Too often this is probably the only advice that can be expected, even in the enlightened society of today. The end result either puts more money in the pockets of the modern charlatans or encourages patients to try self-medication with substances of dubious value. The drugs that are available for treating disorders of sexual function are discussed in detail in Chapter 8.

Testicular Transplants: the Monkey Gland Story

The work of Berthold and others demonstrating that it was possible to reverse the effects of castration by implanting extracts of testes led naturally to the consideration of the possible use of such techniques in man. The early pioneer in this branch of medicine was the Anglo-

French physiologist, Brown-Sequard. In 1889, when he was 72 years old and beginning to experience a diminution of his physical powers, he tried administering testicular extracts prepared from guinea-pigs and dogs to himself by subcutaneous injection in an attempt to reverse this process.[1] He reported that he subsequently experienced a general improvement in his health and vigour. His paper, published in the *Lancet*, in which he describes these experiments on himself, provides some fascinating insights into the beliefs that were held at the time concerning the sexual functioning of the male. He wrote, for example:

'It is known that well-organised men, especially from twenty to thirty-five years of age, who remain absolutely free from sexual intercourse or any other causes of expenditure of seminal fluid, are in a great state of excitement, giving them a great, though abnormal, physical and mental activity.'

A colleague of Brown-Sequard, Dr Variot, also tried injecting aqueous testicular extracts, not in himself, but in three old male patients. All three subsequently reported increased stamina and physical strength. They had all, however, been informed by Dr Variot that they were to receive a fortifying injection, although its nature was not disclosed, and it is possible that autosuggestion could have played a part in the effects observed.

Testicular extracts, prepared according to the method of Brown-Sequard, have since been made and found to contain no detectable androgenic or male hormone activity. The cause of the fortifying effects he experienced thus remain something of a mystery. Nevertheless, an interest in the possible rejuvenating effects of this tissue persisted, and Brown-Sequard was only the first of a number of investigators to seek the new philosopher's stone.

In Vienna, Dr Steinach began a series of experiments using rats. His work, carried out over a period of about 30 years from 1894, led to several interesting discoveries, most notably that the ligation or cutting of the sperm duct (vasectomy) led to an increased life span and prolonged sexual vigour in his rats. Steinach's principal belief was that the symptoms of old age – loss of mental and physical vigour and decline of sexual desire and potency, were largely due to the deterioration of the testes as they became exhausted. The object of his experiments was not specifically aimed at restoring the sexual vigour of the animals, but at overcoming or reversing the degenerative changes associated with old age. In the original experiments with rats, sexual vigour and life span were the only two measures available for assessing any beneficial effects of the surgical procedures. At that time there were no reliable techniques for testing the mental abilities of the animals. Such is the public obsession concerning sexual matters that it was only this peripheral aspect of the research which achieved widespread publicity.

The technique of vasoligature was used by several surgeons to treat ageing patients with enlarged prostate glands, and occasionally there were instances of general improvements in their physical condition. When the operation of prostectomy was perfected, this was the preferred treatment, and again, occasional cases of slight 'rejuvenation' were recorded. The variable success rate with vasectomy as a means of reversing senility, and the uncertainty of the influence of suggestion on the patients, have led to the abandonment of the technique. Vasectomy is now, of course, a relatively common operation among men of younger age groups who wish to limit the size of their families. Instances of increased libido under these circumstances have been put down to the removal of the fear of causing pregnancy. A similar phenomenon has been noted with some women who start taking oral contraceptives (see page 161).

One of the problems besetting the proponents of the use of testicular extracts was the uncertainty concerning the concentration of the active principles in each extract, especially when their chemical identity and stability were unknown. Also, the effects tended to be short-lived when, in fact, a permanent improvement was desired. Voronoff was not the first to attempt testicular transplants in human subjects, but he has become the most well-known name in that he used testicles obtained from various species of monkey for his operations. He also tended to make the most extravagant claims concerning the rejuvenating effects of these transplants.

The first person to transplant human testicular material from one patient to another was Lespinasse, in 1911. The donor was a normal subject and the recipient a man of 38 who had lost both testicles in an accident. The operation was apparently successful. This success encouraged others to attempt the same procedure. One of the most prolific exponents of the practice was a Dr Stanley of San Quentin who had the advantage of a ready supply of material for transplantation from the executed convicts at the local prison. His periodic successes in restoring the virility of patients who had lost their testes either through accident or disease encouraged him to persevere with the technique. A large number of the inmates (656) from the same prison were given

Table 6.3: The outcome of the human testicular transplants performed by L.L. Stanley in California between 1918 and 1921. (The data have been abstracted from the table quoted by N. Haire in his book *Rejuvenation,* published in 1924.)

Symptoms of Recipient	Number of Cases	Number Benefited	Not Benefited
General asthenia (debility)	336	305	31
Sexual lassitude	95	81	14
Rheumatism	58	49	9
Acne vulgaris	66	54	12
Neurasthenia (neuroses involving fatigue)	56	33	23
Poor vision	41	32	9
Asthma	21	18	3
Tuberculosis	17	10	7
Impotence	19	12	7
Epilepsy	5	3	2
Diabetes	4	3	1
Psychopathic inferiority	8	0	8

crude implants of animal testes, cut up into small pieces and injected subcutaneously into the abdominal wall. His results showed a surprising degree of success considering our present knowledge of tissue rejection following tissue transplants. The results are shown in Table 6.3. (It must, however, be remembered that these results date from well before the era of controlled clinical trials, so the apparent cure-all effects of the transplants should be viewed with caution.)

Voronoff, on the other hand, found great difficulty in obtaining human testes for implantation. People were generally unwilling to submit to the voluntary removal of a testicle even after being assured that, having two available, they could easily spare one of them. His tissue grafts, involving material from anthropoid apes transplanted into human subjects, also appeared to be successful, according to his own accounts. Although the impression of the general public was that these transplants were universally successful and could add years of fruitful sexual activity to any man who wished to undergo the operation, Voronoff himself was somewhat more restrained in his list of indications for testicular transplantation. They were: loss of testicles, whether accidental, congenital or surgical; infantilism (i.e. non-development) of the sexual organs; congenital testicular deficiency; senility, whether normal or premature; hardening of the arteries; *dementia praecox* (a relatively unusual form of insanity). The inclusion of senility was probably sufficient to encourage a number of older men to believe that, with some relatively minor surgical help, they could regain the vigour and appetite of their youth while retaining the experience and knowledge of their later years. This is the perennial dream of mankind, to possess simultaneously the benefits of youth and the advantages of experience.

Voronoff provided detailed case histories of many of his patients. A typical example of the type of person seeking this treatment for senility was an Englishman aged 74. His medical history consisted of severe attacks of gonorrhoea, in addition to malaria and smallpox which he had contracted while on government service in India. He was a chronic drunkard. His appearance was of a little old man, bent, obese, with a tired face, dull eyes, a poor memory and a dull intellect; he had been completely impotent for the previous 12 years.

Following his transplant operation, his condition apparently improved considerably. His body became erect and supple, he grew a head of thick white hair and was in a condition to go mountain climbing in Switzerland. Unfortunately he died from acute alcoholic poisoning two-and-a-half years after his operation.

We still have an image of the gorilla as a symbol of size, strength and sexual potency, and the thought of gaining just a fraction of that power by means of a transplant would be a powerful incentive for a physical improvement in condition based entirely on expectation. It is not recorded whether Voronoff's patients were informed that the material they received was most frequently obtained from chimpanzees.

Such was the widespread interest in the claims made by Voronoff, among both the medical profession and the general public, that an international delegation of endocrinologists was sent to visit his clinic in Algiers with a view to assessing the scientific merit of his work. The report sent subsequently by the British contingent to the Ministry of Agriculture was less than enthusiastic. From the recent developments in the techniques of transplant surgery, we now know that it is extremely unlikely that a foreign piece of tissue would survive and be functionally active under the conditions used by Voronoff. The presence of testosterone in the transplanted material might have had a temporary effect, but this would soon have ceased as the tissue was rejected by the body. The isolation and purification of testosterone shortly after this time made possible the much simpler, and safer, injection of pure hormone in cases of loss of testicular function. Testosterone itself was found to have beneficial effects in cases of male senility and could act to restore a diminished libido. The transplantation of testicular tissue from animals was therefore rendered completely obsolete.

The Therapeutic Role of Testosterone

Testosterone administered orally has little effect since, in spite of being rapidly absorbed into the bloodstream from the gut, it is mostly metabolized during its first passage through the liver before it can enter the general circulation. Attempts have been made, with some success, to take advantage of the fact that testosterone can be absorbed through the buccal mucosa of the mouth. Methyl testosterone tablets containing between 5 and 25 mg of the steroid are available on prescription, and are effective when dissolved under the tongue in cases in which only low supplementary doses of testosterone are required. Testosterone undecanoate is available in orally effective capsules. The more usual form of treatment for complete replacement therapy is by depot injection or implants. The hormone or its analogue is combined with a fatty acid to form an ester which dissolves in oil. This will be absorbed more slowly from the site of injection and so a single injection will be effective for a longer period of time. In full replacement therapy the aim is to administer 10 mg of testosterone per day, and this can be achieved by three injections of 25 mg of the propionate or cypionate ester per week. The therapeutic principles are the same as those employed in the treatment of diabetics with insulin injections. Testosterone implants have a very slow sustained release and are effective for between 7 and 8 months.

Testosterone and Male Impotence

In cases of impotence without hypogonadism, testosterone is ineffective as a treatment for the impotence. In hypogonadism, where there is no appearance of puberty or sexual development, testosterone implants have been used with some success, but replacement therapy in these situations can rarely produce a normal sexual development. At the other end of the spectrum there is still considerable debate concerning the value of testosterone treatment for ageing men who are experiencing a gradual, and entirely natural, decline in testicular function. The term 'climacteric' has been employed to describe this decline, but it is not as rapid as the cessation of ovarian activity associated with the menopause in women. Also, it does not seem to cause the same acute physical and emotional problems that women can experience at this time. In fact, the entire concept of the male climacteric is still controversial.

There is no evidence that additional testosterone supplements can increase sexual activity or the libido above normal levels in cases in which the endogenous testosterone level is within the expected range for any particular age group. This is certainly due in part to the feedback controls exerted over testosterone synthesis and release by the gonadotrophins. The introduction of additional testosterone into the system will merely cause suppression of the release of endogenous testosterone.

But what happens in situations where the testosterone level is abnormally low? There is some evidence that male homosexuality is associated with marked differences in the circulating hormone levels in the blood, compared with heterosexual controls. In one study[2] it was found that young males (student volunteers) who were exclusively homosexual had plasma testosterone levels that were less than half of those measured in a corresponding group of heterosexuals, whereas their luteinizing hormone (LH) levels were five times greater than those of the heterosexuals. (The follicle stimulating hormone levels were similar in both groups.) However, the subjects were also regular drug users, and this may have had an influence on their hormone levels. In spite of their low testosterone levels, the homosexual subjects appeared to experience no difficulty in achieving erections and orgasm, and their libido was no different from that of heterosexuals of the same age group. This suggests that, irrespective of their sexual proclivities, the level of testosterone alone had little influence on the libido.

The treatment of homosexuals with androgens such as testosterone in order to counteract the 'feminine' aspects of their personality does not alter their sexual tendencies; and the past use of anti-androgens to suppress their 'deviant' libido did not alter the direction of their interests when their libido returned after cessation of the treatment. Although there does seem to be an endocrine dysfunction associated with some instances of homosexuality, the cause–effect relationship has not been convincingly demonstrated. It is probably naive in the extreme to imagine that the use of a single hormone could alter a condition which has a considerable psychological component.

As mentioned previously, male impotence can be roughly divided into that of psychogenic origin, which often responds to counselling, and constitutional impotence, in which androgen therapy may play a part. Impotence occurs in between 37 and 55 per cent of male diabetics, and can manifest itself as a decline in libido or an inability to ejaculate, but the diabetics' most frequent complaint is that they are unable to achieve an erection. Testosterone levels are usually normal in these patients, as is their response to human chorionic gonado-trophin (HCG), and treatment with either androgens or gonadotrophins is not usually successful. The impotence may well be produced as a consequence of the nerve damage that is associated with juvenile diabetes.

Evidence obtained from patients with either psychogenic or con-stitutional impotence[3] indicated that urinary testosterone levels (which provide a rough indication of the rate of testosterone production) were significantly higher in the psychogenic cases compared with the con-stitutional cases, and were, in many instances, within the normal expected range. Other work has shown that testosterone levels are reduced in sexually abstinent men, but that they rise when intercourse is resumed. This suggests that it is not necessarily the testosterone which is regulating the level of sexual activity and the libido, but rather the reverse. This is a logical relationship from a biological point of view, since testosterone is largely responsible for ensuring the production of mature sperm, and this is only strictly necessary when an individual is sexually active. Clinical trials of preparations containing testosterone for the treatment of male impotence are described in Chapter 8.

Androgen Therapy for Female Frigidity

The sexual drive or libido of women is largely determined by psycho-logical and social factors, and any effects produced as a result of altera-tions in hormonal balance are outweighed by these emotional compo-nents. That is not to say that hormonal balance has no effect on mood and behaviour; premenstrual tension and post-partum depression (the 'baby blues') are both due to temporary changes in the hormonal balance.

Frigidity is defined in medical dictionaries simply as sexual coldness, but there is clearly more to the subject than that. The term frigidity can be applied to a disorder having a psychological or physical origin, and it is used to describe either a lack of interest in sex, or an inability to have an orgasm. The lack of interest in sex may reflect the nature of the personality, or may result from unpleasant sexual experiences in the past. Frigidity often occurs after violent rape, for example, and careful psychiatric counselling is often necessary if a normal sex life is to be regained. In the treatment of frigidity the first step should always be to determine the root cause of the problem.

Figure 6.3: Plasma Levels of Progesterone (—·—·—·—), Luteinizing
Hormone, LH (———), and Total Oestrogens (————), Correlated with
the Subjective Self-assessment of Arousal During the Menstrual Cycle in
Normal Women Not Taking Oral Contraceptives. Because of slight
variations in the length of individuals' cycles, day zero has been
arbitrarily fixed as the day of the LH peak. Ovulation would occur at
about this time. The data have been compiled from Moghissi, K.S.
et al. (1972) *Am. J. Obstet. Gynec.* **114**, 405, and Englander-Golden,
P. *et al.* (1980) *J. Human Stress* **6**, 42. There is little evidence from
these data that sexual arousal can be correlated with the plasma level of
any single sex hormone

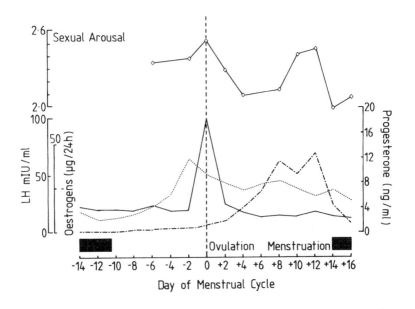

The basis of sexual pleasure in women resides in the clitoris, which
is extremely sensitive to physical stimulation. The sensitivity of the
clitoris is influenced by androgen secretions from the ovaries. This
secretion varies during the normal menstrual cycle and may contribute
to the variation in libido experienced during the cycle. The endocrinological changes that occur during the cycle are summarized in
Figure 6.3. Occasionally, following the surgical removal of the ovaries,
or after long-term oral contraceptive therapy which suppresses ovarian
secretions, the clitoris becomes less sensitive to physical stimulation

and an orgasm becomes much more difficult to achieve. Under these circumstances, androgen therapy may be used to increase the sensitivity of the clitoris and thus relieve the frigidity.

Great care has to be taken in prescribing androgens over long periods of time since virilization can occur; in particular, a deepening of the voice and an increase in facial hair. Some of these changes are irreversible even after the treatment has ceased. The value of androgen therapy (usually testosterone propionate) in the specific treatment of frigidity was discovered quite by accident: women being treated with androgens for other gynaecological problems began to report an increased libido and sexual desire, and this led to an investigation in which the effects of testosterone on the libido were assessed.[4]

The survey was carried out on 101 women who were suffering from either primary or secondary frigidity, as well as control subjects who required androgen therapy for other disorders such as menorrhagia (excessively prolonged periods). No less than 88 out of the 101 women in the study reported increased libido, and of these, 20 reported an excessive increase in clitoral sensitivity. The women who did not respond to the treatment were mostly suffering from secondary or psychogenic frigidity. Among those who did respond to the treatment, several were not able to achieve improved sexual gratification or pleasure because the androgens tended to suppress the normal effects of oestrogens on the vaginal tissues. There was therefore a marked reduction in the normal vaginal lubrication, making intercourse more difficult and painful. The authors of the study concluded that, under appropriate circumstances, androgen therapy could impart the following benefits: increase susceptibility to psychosexual stimulation; increase the sensitivity of the external genitalia, and the clitoris in particular; in most cases increase the intensity of sexual gratification.

However, there are several criticisms that can be levelled at this work before these conclusions are unequivocally accepted. First, there is the difficulty of assessing libido, as opposed to merely recording the frequency of sexual activity. The role of the sexual partner should always be considered in these studies. In the absence of a partner it is important to record the frequency of masturbation as a means of obtaining sexual gratification. Secondly, the study did not include a placebo preparation as a control. The use of a placebo in a double-blind clinical trial (see page 169) is essential in studies where extremely subjective responses are being measured. Thirdly, all the subjects were suffering from some endocrine disorder. In other words, the control subjects were not really controls at all. The presence of menorrhagia may well lead to a loss of libido, and the effectiveness of the androgen treatment would be purely as a consequence of the cure of the menorrhagia. It is therefore very difficult to quantify the effects of androgens on the libido, but it certainly appears to be an avenue worth exploring in the search for a clinically useful aphrodisiac in the treatment of frigidity.

More recent and thorough studies have been carried out in which circulating hormone levels, mood, sexuality and sexual activity have been correlated throughout the menstrual cycle. Testosterone levels in women with normal ovulatory cycles rise slightly during the middle of the cycle (the days around the time of ovulation), and in one study of 11 couples, this was found to correlate with an increase in frequency of sexual intercourse.[15] This was not confirmed in a later study of 55 women,[16] although there was a correlation between average testosterone levels and masturbatory frequency. These two results confirm the findings of Schreiner-Engel and her colleagues who found that vaginal responsiveness to erotic stimuli was highest in the women with the highest testosterone levels, but that these same women tended to report having less satisfactory sexual relationships with their partners.[17] It seems that androgens in women may positively affect only autosexuality and have negligible or even negative effects on heterosexual activity.

Prolactin

Prolactin is a peptide hormone closely related to human growth hormone which is responsible for the initiation and maintenance of lactation following the birth of a baby. It undoubtedly has many other functions in the body which are not necessarily related to reproduction. The possible influence of prolactin on the libido has been implied from various indirect pieces of evidence: for example, male patients on maintenance haemodialysis treatment for kidney failure often suffer from impaired sexual function as a consequence of hormonal as well as psychological factors. Plasma prolactin concentrations have been found to be abnormally high in these patients, and this high level has been associated with the reduced libido and sexual activity.

Bromocriptine is a drug which blocks the release of prolactin and it has been used successfully in the treatment of male impotence in haemodialysis patients.[5] Twenty male patients were given bromocriptine on a single-blind crossover trial against a placebo (see page 226 for details of the experimental design). Sexual function was evaluated by giving a questionnaire. Some patients suffered from a severe drop in their blood pressure as a result of the drug therapy and were obliged to withdraw from the trial. In the seven patients who completed the eight-week course of treatment, six reported increased libido during the bromocriptine treatment and the same number enjoyed an increased frequency of sexual intercourse. The placebo was without effect.

Unfortunately, in addition to the hypotension mentioned above, bromocriptine had other side-effects; including nausea (7 out of 15), vomiting (4 out of 15), double vision (1 out of 15), and loss of appetite (8 out of 15). This very high incidence of unpleasant side-effects probably makes bromocriptine unacceptable from a safety point of view for the treatment of impotence. However, the discovery of a new anti-prolactin agent lacking these adverse actions might prove to be an extremely useful drug in the treatment of male impotence. The scientific rationale for the use of bromocriptine and other related drugs in various neurological disorders is described in Chapter 8.

Oral Contraceptives and Frigidity

There have been a very large number of case reports published in the medical literature in which changes in the sex drive have occurred following the administration of oral contraceptives over various lengths of time. Among women with a 'normal' libido, and otherwise satisfactory sex life, the removal of the fear of an unwanted pregnancy can encourage sexual activity, and it is still thought that the pill has been a significant factor in the recent increase in promiscuity, particularly among the young. On the other hand, some women find, perhaps subconsciously, that sex becomes less appealing when there is no possibility of becoming pregnant. There is no doubt that the pill can have significant psychological effects upon the libido, quite apart from its established endocrinological effects. These effects can be of a positive or a negative nature.

It is known that oestrogen deficiency, which can occur in menopausal women, often leads to a loss of sexual interest which can be restored by oestrogen replacement therapy. A pill containing relatively high levels of oestrogen could act by this means to increase the libido. In contrast, the progesterone in the pill is capable of suppressing androgen secretion by the ovaries, and this would result in a loss of clitoral sensitivity. Androgens are also secreted by the adrenal glands (removal of the adrenals almost invariably causes a loss of the sex drive), but in some cases this is insufficient on its own to sensitize the clitoris. When this situation arises it is possible to substitute a pill containing an alternative progestogen with some androgenic potency.

Some of the lesser side-effects of oral contraceptives may also interfere with the libido indirectly. These include headache, tenderness of the breasts, depression, and reduced vaginal lubrication. If the anaphrodisiac effects are significantly affecting the patient's sex life, it is often possible to find an alternative pill containing different levels of oestrogen and progestogen.

Use of Sex Hormones as Anaphrodisiacs

Over the last fifteen years or so, it has been found that male hypersexuality can be specifically treated with anti-androgens: steroid-based drugs which prevent the endogenous androgens (particularly testosterone) from exerting their normal actions within the body. Both oestrogens and progestogens have anti-androgenic effects (see above), and this may be one of their natural functions in the male. However, the term anti-androgen is now usually reserved for chemical analogues of testosterone, such as cyproterone, which compete directly with testosterone at its sites of action. In this respect cyproterone is 250 times more potent than progesterone.

Assessing the effects of anti-androgens on the libido in animal tests, prior to their introduction into clinical usage, has proved to be difficult. There are very large variations between species in the influence of a loss of natural androgenic action on the libido. For example, both dogs and monkeys can maintain a normal level of sexual activity for up to four years following castration, although their fertility is lost. Bearing in mind the relatively shorter life span of these species compared with man, this is quite remarkable. It is not therefore surprising that the anti-androgens themselves have little effect on the libido and sexual activity of laboratory animals.

Cyproterone acetate in the human male has none of the unpleasant feminizing side-effects of oestrogens, but it still suppresses the libido and reduces the sexual potency within about two weeks of commencing treatment.[6] The physiological effects of the drug are first manifested by delayed ejaculation, followed by an increasing inability to ejaculate. There is also loss of fertility, and the sperm count can even fall to zero. The circulating testosterone level stabilizes at about 20 to 30 per cent of normal, although this is not the only basis of the anti-androgenic action of the drug. The effects are all reversible, and normal sexual function returns shortly after ceasing the treatment.

The principal use of cyproterone is to reduce the libido in males who show a deviant or aggressive hypersexuality which can lead to criminal offences being committed. The moral justification for the use of such drugs in the treatment of sex offenders has been questioned, but it is worth noting that many prisoners, following their conviction for sexual offences, volunteer for this form of treatment as a means of reducing their abnormal or excessive urges.

Although androgens have been used in women to increase the sexual drive, the use of anti-androgens is restricted to the treatment of precocious puberty. In this situation the drug stops menstruation and reduces the size of the breasts. In both boys and girls suffering from adolescent acne, anti-androgens can relieve their condition, which results from the increased sebaceous activity of the exocrine glands in the skin. Excessive androgen secretion by the ovaries or adrenals can cause women to undergo masculinization, particularly with regard to facial hair. This condition has also been treated with anti-androgens. In normal adult women with a tendency to hirsutism or acne, as a result of the over-production of androgens, a contraceptive pill has been produced which contains an anti-androgen. The effects of anti-androgens on the female libido have not yet been thoroughly investigated, although a detrimental effect might be expected.

It is worth considering how folk knowledge is often found to have a sound scientific basis by the results of seemingly unrelated research. It is a well-established belief in many countries that women with pronounced body and facial hair are particularly sexy and have a high libido. This belief may have some foundation, in that excessive body hair suggests a high level of androgen secretion which, as we already know, tends to increase the libido. It is interesting that in North America and Northern Europe, women tend to shave their arms, legs, armpits, and even the pubis as a cosmetic procedure whereas, in the mediterranean countries, body hair is often retained and regarded as a sexual attractant.

A not dissimilar myth attaches to the sexual capacity of bald men. There is still no scientific explanation for the apparently paradoxical effects of androgens on hair growth in the male. The increase in male sex hormone levels at puberty is associated with the stimulation of hair growth over large areas of the body, yet it also eventually leads to a gradual atrophy of the hair follicles over the temples and on the crown of the head. This baldness is a sex-dependent dominant generic trait, but it is not, as is generally believed, a sex-linked phenomenon. That is, in women both parents must possess the necessary gene in order for baldness to occur, but in the male only one parent is required to carry the gene in order for it to be expressed as baldness in the offspring. As a consequence of this inherited trait, about 75 per cent of European males over the age of 50 suffer from baldness. In some cases baldness may occur much earlier in life, and in some races it does not occur at all.

It is an odd contradiction which encourages bald men to buy vast quantities of quack remedies, or undergo expensive cosmetic surgery, in order to restore their male ego by exhibiting a full head of hair when, in fact, the opposite state seems to be associated with virility, at least in the minds of many members of the opposite sex. It has been reported that the prolonged administration of anti-androgens has resulted in the regrowth of hair on a bald head, but the contingent symptoms of loss of libido and sexual potency would seem to defeat the aim of the treatment if only masculine pride is at stake. There should be no reason for a man to feel ashamed or inadequate as a consequence of his state of baldness, so long as a sufficient number of women are attracted to this condition in the belief that it is a sign of a high sex drive. Unfortunately, no similarly encouraging folk belief can be invoked in the case of bald women.

Assessing the Effects of Potential Aphrodisiacs

The physical manifestations of sexual arousal are, particularly in the male, fairly easy to quantify. Measures such as the size and number of erections, number of orgasms, and frequency of sexual intercourse over specified periods of time enable a reasonably accurate estimate of sexual activity to be made (that is, assuming of course that the subjects under study are telling the truth). Using parameters such as these, it has been possible to detect the positive effects of various drugs which are currently prescribed for their aphrodisiac properties. Penile erections can be measured under laboratory conditions by the use of a pressure transducer which converts the volume change produced by the swelling of the penis into readily quantifiable electrical signals. By this means the effects of watching erotic films or reading erotic books on one particular measure of the level of male arousal have been investigated.[9] Not surprisingly, perhaps, the authors concluded that erotic material was capable of evoking the physical symptoms of sexual arousal.

Physiological indications of the level of sexual arousal in women are more difficult to measure, but the increase of blood flow through the labia and vaginal wall, which occurs during sexual arousal and is analogous to the increased blood flow into the penis in the male, does lend itself to objective measurement. Changes in the blood flow through a tissue or organ produce changes in its surface temperature, and this can be measured by an electronic thermometer. The vaginal blood volume has been measured using a plethysmograph coupled to a light source and a photosensor device.[8] Later studies have indicated that it is more accurate to record a number of haemodynamic measures such as heart rate and vaginal pulse in order to assess the occurrence and intensity of orgasm.[14]

In one typical study,[8] eight women volunteers had their labial temperature and vaginal blood volume recorded while they watched an explicitly erotic film. Control recordings were made before and after the showing of the film. While the film was being viewed, the labial temperature increased significantly by between 0.5 and 1.2°C, and then remained elevated for some time after the film had ended. The vaginal blood flow also increased, although this was not statistically significant or consistent.

The subjects were asked, in addition, to assess their feelings of genital arousal on a 7-point scale:[9] 1, no genital sensations; 2, mild genital sensations; 3, moderate genital sensations; 4, slightly strong genital sensations; 5, strong genital sensations; 6, vaginal lubrication; 7, orgasm.

The increase in labial temperature correlated very well with the subjective self-assessments, although individual values varied widely. It would appear, therefore, that a relatively simple physiological measure can be used to detect and quantify female arousal. This would be very valuable in determining the effectiveness of a potential aphrodisiac that was claimed to act directly to increase the level of arousal in response to any given erotic stimulus.

Psychological assessments are far more difficult to make, and are open to considerable uncertainty since one is dealing with the mood and thought processes of an individual. Psychiatric rating scales are well established within the context of anxiety or depression, but sexuality rating scales are more controversial. They are based largely on the observation of psychiatric patients by clinicians and are really designed to assess the degree of mental disorders which can be reflected in extreme or abnormal sexual behaviour. One such rating scale[7] is shown below:

Score:	+6	+5	+4	+3	+2	+1
Sexuality:	Assertive ...	Soliciting ...		Overactive ...	Baseline	

Score:	−1	−2	−3	−4	−5	−6
Sexuality:	Passive ...	Homosexual ...		Homosexual (assaultive)		

Patients are initially placed by the psychiatrist either to the left (positive, active heterosexual) or to the right (negative, passive or latent homosexual) of the middle line. The same psychiatrist then scores the patient's behaviour during the period of treatment. This particular scale is only one out of a total rating which takes into account 19 functions of behaviour. It is clear that this scale has been designed for assessing disturbed individuals rather than the 'normal' spectrum of sexual behaviour. The same type of scoring, however, has been used in some of the studies where drugs, which have been prescribed to psychiatric or geriatric patients, have been found to possess unexpectedly aphrodisiac side-effects. Specific examples are L-DOPA and *l*-tryptophan, which are discussed later (see Chapter 8).

Scores of this type provide no guide to the subject's sexual feelings except in so far as they are expressed in overt sexual behaviour. To obtain information at this level, self-assessment provides the best alternative. A very simple rating scale which can be adapted for this use involves a written questionaire which has, after each question, a horizontal line 10 cm long. For example:

Q. How sexually receptive do you feel at present?
Completely unreceptive . . . . Most receptive ever

---

The subject is required to make a mark on the line at the point which he feels most closely corresponds to his mood or feelings at the time. A mark in the exact centre of the line might be regarded as a normal baseline level of receptiveness for that individual. The investigator now measures the distance between the left-hand end of the line and the subject's mark. This provides a quantitative score for that question. Clearly, the choice of questions will be all important, but the main advantage of this system is that the usual variation between individuals is largely eliminated. For example, taking the sample question given above, different subjects might well have widely varying baselines of sexual receptiveness, but they will all be normalized by the arbitrary fixing of their individual baselines to a value of 5.0 cm. From this reference point, any positive or negative change can be readily detected.

This technique has been applied successively to the clinical testing of sleeping pills where volunteers are required to assess the quality of their sleep and morning performance. The point behind the discussion of this rather detailed methodology is to illustrate that it would not be impossible to detect, scientifically and with some degree of precision, an aphrodisiac effect that was manifested either in terms of physiological or psychological effects in either sex. It would be very interesting to submit some of the current aphrodisiac preparations to this sort of investigation in a controlled double-blind placebo trial.

The Placebo

'Placebo Domino in regione vivorum'[10]

A placebo was originally a professional mourner who was paid a fee to sing at funerals, thus relieving the deceased's relatives of the task. The medical definitions of placebo evolved at the end of the eighteenth century and referred to medicines that were administered solely in order to satisfy the patient rather than render him any particular benefit, but by the end of the nineteenth century the word had become widely understood to mean an inert or dummy medicine.

Many early remedies were undoubtedly placebo in their action, and with a few spontaneous remissions occurring at fortuitous times, together with the positive expectations of the patient, a 'drug' could rapidly acquire a considerable reputation. Since the advent of the necessity for testing new drugs in limited clinical trials prior to their release on to the market, the placebo has become an important feature of controlled clinical trials. It is not enough to compare a new drug treatment with no treatment at all; the act of taking a drug in itself has a psychological effect depending upon the level of expectation of the patient. It is to eliminate this psychological component of a possible therapeutic action that inert placebos, indistinguishable from the active preparation, are included in the trial.

The therapeutic power of the placebo is amply demonstrated by the variety of ailments that respond to placebo therapy (see Table 6.4). The table only reflects a small proportion of the disorders which have been

Table 6.4: Incidence of Placebo Responses (data from Haas *et al*.[11])

Symptom	Number of Studies	Number of Patients	Number Responding to Placebo
Pain	25	961	269
Headache	9	4588	2844
Neurosis	6	135	46
Common cold	3	246	110
Epilepsy	1	72	0
Asthma	2	19	1
Constipation	3	144	17

investigated for placebo responses. Some ailments are clearly more likely to respond to placebo therapy, and the results from the 4588 patients with headaches implies that one has a slightly better than even chance of getting relief from a chalk pill. Those disorders which merely represent the overt symptoms of some underlying physiological or psychiatric problem seem to be the most susceptible to suggestion, whereas those that have a more defined pathological basis are far more recalcitrant to placebo therapy. Many disorders of sexual behaviour belong in the first category, and the placebo effect plays an important part in the action of most aphrodisiacs, both past and present.

Drugs are not the only things that can exert a placebo effect: surgical and other diagnostic procedures have also been found to share this property. Some recent evidence has suggested that even electro-convulsive shock therapy for the treatment of depression can some-times be effective when the equipment is not switched on and no shock is delivered to the patient. Evidence of this type has led to the belief that, within the population, there are a significant minority of so-called 'placebo reactors' who will respond to procedures or treatments that possess no intrinsic efficacy. This phenomenon would explain the continuing success of quack doctors and alternative medicine.

In the case of sexual performance, where the physical and mental state are equally important in determining the degree of inclination, appetite or ability, a marked placebo response is almost bound to occur. Apart from the importance of the personality of the individual in determining the likelihood of a placebo response, the doctor's personality and attitude are also critical. If the patient is anticipating that a prescribed medicine will be effective, then the conditioned belief or expectation in the efficacy of the medication will have a significant influence on the outcome of the treatment. A sympathetic and calm doctor who believes in the value of the drugs he is prescribing, and who communicates this feeling to the patient, is more likely to achieve a successful cure than a short-tempered cynic who takes little personal interest in the patient. This last factor is well summed up in the term 'bedside manner': a technique which need not be restricted to the bedroom.

Placebos can also produce negative reactions which resemble the side-effects associated with active drugs. Some people are unusually suggestible, and it is not difficult for them to imagine that they are taking active drugs and they respond accordingly, even though all they have been given is a placebo. Commonly reported symptoms have included headache, nervousness, drowsiness, insomnia, and nausea. In one case, a schizophrenic patient became physically addicted to the placebos with which she was being (inadvisably) treated, and she showed all the classic symptoms of a withdrawal syndrome when the placebo treatment was stopped.

The physical properties of the placebo are extremely important. It would be of little use to provide a sweet-tasting capsule as a control against some bitter-tasting active drug. Similarly, the colour of tablets or capsules is highly influential in conditioning the patient to the expected therapeutic effect. Sleeping pills are often blue or green, and never red or orange. In the case of vitamin pills and commercial aphrodisiac preparations, the reverse is true, red and orange being the colours associated with stimulation, excitation and power.

The Double-Blind Crossover Clinical Trial

This is the most common design for clinical trials of potentially useful new drugs. The essence of the scheme is that a population of patients or healthy volunteers, as nearly homogeneous as possible in terms of age, sex, health, etc., are divided randomly into two equal groups. One group is given the active drug and the other is given the placebo. After a suitable period of time the two groups are switched, so that the first group now has the placebo and the second the active drug. This is the crossover principle, which enables each subject in the trial to act as their own control. Neither the subjects nor the immediate medical staff administering the drugs should be aware of which group is which; in fact, the drugs are usually labelled in code. This is the double-blind principle which is intended to eliminate any psychological influences. At the end of the trial the solution to the code is revealed and the improvement achieved by the test drug over and above that produced by the placebo is assessed by a statistician. At least, this is the ideal;

in practice it is occasionally found that some flaw in the design of the trial has made the interpretation of the results difficult or even impossible.

The controlled clinical trial of an aphrodisiac is fraught with difficulties. Although there might be no problem in obtaining healthy volunteers for such a trial, it might be more appropriate to test the drug in cases where there is some underlying sexual problem. It would not be easy to find a homogeneous population of patients with similar sexual problems, and the number of patients available might not be sufficient to enable a satisfactory statistical analysis to be carried out. The few examples of aphrodisiac trials in the medical literature amply demonstrate the difficulties of interpreting results, and all have been criticised by subsequent investigators.

Placebos and Drug Abuse

Drug addicts are generally very good placebo responders; the self-administration of drugs of abuse is a situation where the level of expectation of psychotomimetic effects is very high, and the behavioural effects of the drugs are consequently due in part to non-pharmacological factors. One group of researchers were able to induce symptoms of moist skin, headache, fatigue, drowsiness, anxiety, chilliness, dreamy feelings, increased appetite, unsteadiness, hotness, weakness, and heavy feelings in the hands and feet, in between 25 and 60 per cent of volunteers given 75 ml of ordinary tap water in the belief that it contained LSD. But, on the other hand, none of the volunteers experienced the visual hallucinations that would be expected after a dose of LSD. Unlike the symptoms listed above, it is extremely difficult to induce hallucinations purely on the strength of will-power. Regular drug users seem to learn the appropriate responses to the drugs they are taking in order to expand their minds.

For a short time at least, dried banana skins were reputed to produce psychedelic experiences when smoked. Subsequent controlled experiments showed that the dried skins had not the slightest hallucinogenic effect, and the fashion for smoking bananas soon passed. There was a scientific, albeit remote, rationale for this belief in that banana skins do contain relatively high concentrations of two physiologically active amines: serotonin (5-HT) and tyramine. The popular hallucinogen LSD is known to interact with the 5-HT system in the brain, although it is not known whether this is the basis for its hallucinogenic properties. Another brief myth concerned the seeds of morning glory (*Ipomoea*). For a short time it was not possible to buy these seeds over the counter in case they did contain some hitherto unknown active principle, but good sense eventually prevailed and they are now legally acceptable to the authorities.

These two examples serve to illustrate the problem that is inherent in assessing the efficacy of aphrodisiacs. The reputation of any drug with behavioural effects involving either sexual stimulation or psychedelic actions depends very much on the inherently unreliable reports of the users and distributors of the drug. In the case of such a complex and subjective response as might be expected from an aphrodisiac, the expectation of the user must play an important part in the overall effect experienced.

Placebos as Aphrodisiacs

Of the host of substances which throughout history have been hailed as aphrodisiacs, the vast majority are active only in the sense of a placebo effect. The willingness of people to have faith in bizarre or expensive products has hardly diminished over several thousand years. With the benefit of hindsight we can now regard the use of animal excrement or mystic incantations as either disgusting or superstitious in turn, but the advent of our present age of scientific enlightenment has not completely overcome the inherent wish to believe. New megavitamins or superhormones of unspecified nature have taken the place of lizards' hearts and cocks' testicles, and it seems the constant public demand for sexual remedies will always be present to help line the pockets of the unscrupulous.

It was mentioned earlier that the appearance of a placebo was an important factor in determining its potency, and this aspect has not been neglected in the preparation of many modern aphrodisiacs. The colours red, orange, yellow and gold predominate. Chemical analyses of various proprietary remedies has shown that the yellow colour is usually derived from turmeric, and the red from paprika and cayenne pepper. All these spices have reputations as aphrodisiacs in their own right. The next important principle for achieving a placebo effect is to ensure an appropriate taste in the preparation. For an aphrodisiac, a 'hot' or burning sensation is ideal and will match the expectations of the consumer. Some chilli pepper or cayenne is excellent for this purpose; the 'potency' of the mixture being adjustable according to the amount of chilli added. The powerful rubefacient (irritant) action of chilli pepper is maintained throughout the length of the gastro-intestinal tract and, eight to twelve hours after the consumption of a particularly hot curry, the effects of the chilli will be detected at the opposite end of the digestive system, with occasionally painful results. It is fortunate that, in contrast to cantharidin, the irritant and blistering principle (capsaicin) present in hot peppers is not absorbed into the bloodstream or excreted through the kidneys. If it were, then Madras curries might be expected to provide the same unpleasant symptoms as Spanish flies.

Having achieved an aphrodisiac preparation that both looks and tastes hot, and consists only of normal kitchen spices, the next task is to invent an appropriate name and compose some suitably vague advertising copy that will not attract the unwelcome attentions of the advertising standards authority. (Some fine examples of the art have already been illustrated in Chapter 1.) By including only common culinary ingredients there is, of course, no question of having to submit the preparation to scientific analysis in order to obtain a product licence. In the UK there is a legal, if not entirely scientific, distinction drawn between food and drugs.

The placebo effect has even appeared in erotic fiction. In the story of Lilith by Anais Nin, the husband tells his apparently unresponsive wife that he has substituted Spanish flies for her saccharin tablets. Lilith spends the evening following this revelation in a state of anxiety and anticipation, waiting to be overcome by uncontrollable sexual passion. Unfortunately, she remains unaffected, even after taking another larger dose of the supposed Spanish flies without her husband's knowledge. The next day, on being told by her husband of the deception, she determines to find some active aphrodisiac that will succeed in stimulating her appetite. The moral of the story is probably that the fault lay in the husband, who did not take sufficient trouble to arouse his wife's obviously latent appetite in the first place.

In conclusion, there is a strong case to be made out for the potency of placebos as aphrodisiacs, although their effectiveness will depend very much on the nature of the person using them. An unshakeable belief in the effectiveness of the preparation is essential if the maximum benefit of the placebo response is to be gained. On the other hand, one point in favour of aphrodisiacs of this type is that they are, by definition, inert and therefore unlikely to produce any adverse or dangerous side-effects in the consumer.

7 DRUGS OF ABUSE

'Some men need some killer weed,
and some men need cocaine.
Some men need some cactus juice,
to purify their brain.
Some men need two women,
and some need alcohol.
Everybody needs a little something,
but Lord I need it all.'

Shel Silverstein

Drug abuse is not just a modern phenomenon: it is probably as old as the drugs themselves. It is found in all societies and creates social and legal problems wherever it occurs. Making certain drugs available only on medical prescription does not prevent them getting into the wrong hands. Making the possession of a particular substance a criminal offence does not prevent people from manufacturing and distributing it, and otherwise law-abiding citizens from buying it. Problems arise, therefore, both with controlling illegal substances and with the misuse of legal drugs or other products. The current fashion of solvent abuse by young people in Britain, for example, is made more difficult to control by the widespread availability of these solvents in glues and other household products. It has been argued that there will always be a small proportion of psychologically inadequate people who need to find a magical solution to life, and that this solution often takes the form of drug abuse.

It has been suggested that a society tends to abuse the drugs that produce effects culturally relevant to that society. For example, Western society, which values the qualities of aggression and extroversion, tends to employ alcohol, which releases social inhibitions and unmasks aggression. Eastern cultures, on the other hand, have historically used opium, which produces dreams and introspection, emphasizing the phlegmatic and introverted qualities prized in the East.

In the broadest sense, any non-medical use of drugs might be termed drug abuse, but on the other hand, people often use alcohol, for example, to overcome personality problems such as anxiety or depression; medical conditions for which a doctor might well prescribe an

173

even more potent drug. In this situation the abuse only becomes evident when the individual relies on the drug to maintain a balanced existence. In the extreme case this develops into full-blown addiction or physical dependence. It is typified by a tolerance to the effects of the drug and a marked withdrawal phase, or abstinence syndrome, when the drug is stopped. At a lesser level there occurs psychological dependence, or habituation, in which the drug or even food has become part of normal everyday existence. The drug may be used to excess, but the difference here is that tolerance and withdrawal are not evident, so that it is much easier to stop taking it. There is a theory that tolerance and dependence are responses which are common to all addictive drugs, and that certain members of the population are at risk by virtue of their genetic make-up. However, addiction is also a communicable and social disease that is acquired and maintained through others.

The drugs that are abused tend to be those which are psychoactive and produce changes in mood, perceptions, or the state of consciousness. In fact, although the drugs of abuse have a wide range of pharmacological actions, it is interesting that they can be very broadly divided into two general classes: nervous system depressants, namely alcohol, barbiturates, benzodiazepines, cannabis, methaqualone, opiates (morphine, heroin, etc.) and organic solvents; and nervous system stimulants, namely amphetamines, caffeine, cocaine, nicotine, and alcohol (low doses only).

The people who use drugs recreationally for their psychedelic effects and to escape from the realities of life often also use sex as a means of escaping or expressing defiance of society's standards. It is not therefore surprising that drug abuse and sexual promiscuity are closely related. In the case of the narcotic-drug addict, the sheer cost of financing the habit from illicit sources often makes prostitution or crime the only option. With the so-called softer drugs, their social use is often associated with sexual activity in just the same way that alcohol is used as a means of reducing inhibitions.

Regular drug users, by their own accounts, employ any or all of the substances described in this chapter to enhance their sexual pleasure. Some of this enhancement is probably due to expectation and autosuggestion, but since the state of the mind and the mood are so important in human sexual activity, there is no doubt that these drugs will have significant effects. In some cases they are used to provide an alternative to sex: heroin addicts describe the 'rush' they feel after injecting the drug as being like an orgasm. In spite of the beliefs of the users, there is now increasing evidence that many of these drugs, after prolonged or excessive use, actually have marked detrimental effects on sexual functioning and fertility.

In discussing the nature of aphrodisiacs at the beginning of this book, it was made clear that one should always differentiate between the sexes in assessing the value of a drug as an enhancer of sexual ability. It is also important to consider the essential differences between heterosexual and homosexual practices in relation to drug effects. The Haight–Ashbury Clinic, in the heart of the drug and sex-orientated population of San Francisco, has been in a unique position to record the drug abuses and preferences of its very large clientele. A recent report[1] has listed the sex drugs employed in terms of the sexual predilection of the user. In order of preference they are:

Males:
    Straight: cocaine, cannabis, alcohol, LSD, MDA
    Bisexual: cannabis, nitrites, LSD, cocaine, MDA
    Gay: cannabis, nitrites, MDA, LSD, amphetamine
Females:
    Straight: cannabis, alcohol, methaqualone, cocaine
    Bisexual: nitrites, MDA, cannabis, LSD, methaqualone, cocaine
    Gay: MDA, nitrites, cannabis, methaqualone, cocaine, LSD

Figure 7.1 shows an attempt to quantify the relative preferences of drug users in terms of drug effects on libido and performance.[1] The numbers across the top of the figure merely represent the number of reports from individuals concerning the positive or negative effects of the drugs in question. The sample size is rather small and the reports are purely subjective, so that whereas it is clear that heroin and methadone are almost invariably deleterious, most of the drugs have produced both positive and negative responses and no clear trend is apparent. The two examples which do suggest a possibly real effect are that of Mandrax on the female libido and cannabis on the male libido. There is little scope now for the scientific study of Mandrax following its withdrawal from clinical use, but there have been several investigations into the effects of cannabis on sexual function. The results have been less encouraging than one might be led to believe from the data in Figure 7.1.

The sections that follow should help to provide some explanation of why, for example, the nitrites are used almost entirely by homosexuals, and why methaqualone is only popular among females irrespective

Figure 7.1: Self-reported Effects of Various Drugs on the Libido and Sexual Performance in a Number of Narcotic-drug Addicts. The figures along the horizontal axis represent the actual number of addicts reporting either a positive (+) or negative (−) effect on either function. The data are from Parr, D. (1976) *Br. J. Addict.* **71**, 261. Although the author did not subject his data to statistical analysis, the figure suggests that, with the exception of cannabis in males and Mandrax in females, none of the drugs had any consistently positive effect upon libido or performance

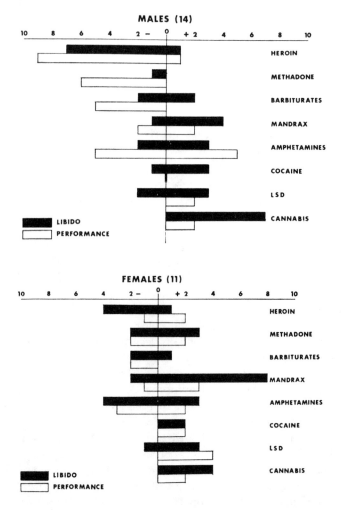

of their sexual persuasion. For the rest of the drugs there is some case to be made out for their aphrodisiac effects, but their promiscuous use is certainly harmful and cannot be recommended or advised.

Amphetamines

Amphetamine itself is *dl-β*-phenylisopropylamine, and is closely related both chemically and pharmacologically to the catecholamines adrenaline and noradrenaline. It is probably better known under its trade names of Dexedrine or Benzedrine. A closely related drug, metamphetamine (*dl-N*-methyl-*β*-phenylisopropylamine), is also described (rather inaccurately) as an amphetamine, as is hydroxyamphetamine. Most of the results described below refer to the drug *d*-amphetamine or dexedrine:

amphetamine   metamphetamine
(Dexedrine)   (Methedrine)

Amphetamine was, for a while, widely prescribed for the treatment of fatigue, both physical and mental, and depression. It is very effective at inducing an elevation of mood and even feelings of euphoria. It was also prescribed as an appetite suppressant in cases of obesity and occasionally as an antidote to overdoses of sedative–hypnotic drugs such as barbiturates. One immediate side-effect that became apparent during clinical use was insomnia, and the information supplied to general practitioners by the manufacturers recommended that amphetamines should not be taken after 4.00 p.m. in order to avoid this complication. Several reports appeared of patients becoming tolerant to the effects of the drug and some instances of patients developing paranoid psychoses as a result of taking it over long periods of time. Any drug that produces euphoria is liable to be abused, and by 1957 amphetamines were placed on Schedule IV of the poisons rules. It was then recognised that the dysphoria and depression that often occurred following the withdrawal of the drug often led to an unwillingness on the part of the patient to give up taking it. There was therefore a strong tendency to habituation or psychological dependence, and in some cases patients actually became physically dependent on the drug.

Immediately following the Second World War, large quantities of amphetamine became available from stocks held by the US and Japanese military and used to suppress fatigue among soldiers in battle. An epidemic of amphetamine addiction occurred as a result of the ready availability of the drug at this time.

Amphetamines are now rarely prescribed, having been replaced by better and safer antidepressants and appetite suppressants. At one time (1959), about 2.5 per cent of all National Health prescriptions in the UK were for amphetamine and amphetamine-containing preparations. However, a flourishing black market in amphetamine still exists, based largely on the illegally synthesized drug rather than thefts from chemists' shops and forged prescription forms.

The main reason for the illicit use of amphetamines is undoubtedly the stimulant properties and the feeling of euphoria they can produce. They are also used by people wishing to remain awake and alert for long periods of time. The question of whether amphetamines have any effect on sexual activity has been neglected by most scientific reviews, and where it is considered, the results are often conflicting, relying much on the anecdotal evidence of individuals. One of the most thorough investigations[2] on this aspect of the action of amphetamine made the important point of assessing the pre-existing sexual adjustment of the subjects in the study (who were all regular amphetamine users). Out of a total of 14 subjects (eight males, six females), 13 could be said to be maladjusted or abnormal in some respect, ranging from a basic frigidity in three of the females to a male who was a homosexual and a pervert. The daily dose of amphetamine consumed by these subjects varied from 30 mg to 1000 mg. (The normal clinical dosage was considered to be 5 to 15 mg per day.) In five cases there was a marked increase in sexual drive and pleasurable sensations. In three of the males, a prolonged delay to ejaculation was reported, which enabled them to perform sexual intercourse or masturbation for hours at a time. Some addicts have used amphetamine specifically for this purpose and it does appear to be a fairly common side-effect at high dose levels. On the other hand, it can be very difficult to obtain an erection in the first place. In a further three cases there was either no change or an increase in feelings of inhibition. The repressed homo-sexual males became more overt in their behaviour and one subject was subsequently prosecuted for exposing himself in public.

It should be added that 11 of the 14 subjects in the study experi-enced psychotic episodes while taking the amphetamine. The authors concluded that the effects of amphetamine were dependent upon the pre-existing sexual state of the individual. On balance there appeared to be an increase in libido, although it often manifested itself in ab-normal or perverse traits.

The prolonged erection without ejaculation in the male has been described by various authors.[3] The average incidence appears to be about 20 per cent, although the dose of amphetamine is obviously important and this is not always known with any degree of certainty. Amphetamines and other illicitly obtained drugs are rarely pure, being 'cut' with inert substances by the dealer in order to increase his profits. Some of the subjects who experienced this phenomenon also stated that it produced feelings of intense frustration when, after several hours, they still could not achieve an orgasm. One subject was driven to mutilate himself in order to relieve the sexual pressure he felt.

In the female there is also a delayed orgasm although the evidence for enhanced sexual pleasure is weak. The inability to achieve an orgasm has often been regarded as one of the prime causes of nympho-mania in women, and is not considered to be a particularly pleasant syndrome. From a woman's point of view, the achievement of frequent multiple orgasms is likely to be a far more satisfying experience.

From a pharmacological standpoint it is interesting to compare the effects of amphetamine with those of cocaine and yohimbine. Amphet-amine and cocaine are very similar in their actions, but yohimbine acts in the opposite way. An ejaculation is triggered by adrenergic nerve impulses, and these are initially potentiated by amphetamine which provokes the release of noradrenaline as well as acting directly to stimulate the receptors. However, overactivity of the adrenergic nervous system can lead to a desensitisation of the adrenergic receptors so that they become unresponsive to further stimulation. The initiation of an erection requires a dilatation of peripheral blood vessels; the effect of amphetamine is to produce vasoconstriction and it is therefore easy to see why amphetamine may make the initiation of an erection more difficult. (The physiology of the erection is described in detail on page 146.)

Amphetamine was usually prescribed in tablet form for oral admini-stration, but its ready solubility makes it ideal, from an addict's point of view, for intravenous injection. Many narcotic addicts have used amphetamine as an alternative when their usual drugs become un- available. Taken in this way amphetamine has the immediate effect of producing a 'rush' similar to that obtained from morphine or heroin and described by one addict as 'like having an orgasm all over your body', although there is apparently an awakening feeling associated with it rather than the drowsiness that occurs with the narcotics. Regular injections (every two to three hours) may keep an addict 'high' or 'speeding' for several days. During this time addicts often show repetitive stereotyped behaviour (known in Scandinavia as *punding*) which can be quite bizarre and inappropriate, as well as paranoid behaviour which is often indistinguishable from schizophrenia. It is during the 'coming down' phase after amphetamine that some users feel so depressed as to be suicidal.

To conclude, amphetamine does have marked effects on the ability of the male to delay ejaculation and the female to delay orgasm. This is the basis for the belief that amphetamine enables users to indulge in sexual marathons. For men suffering from premature ejaculation there are several safer ways of producing a degree of local anaesthesia, and the risks associated with regular use of amphetamine, apart from any legal considerations, hardly justify the experiment. In both sexes, the inability to achieve an orgasm except after several hours' effort may be sufficiently frustrating as to actually decrease the sexual pleasure obtained. It must also be added that the induction of paranoia and unmasking of latent aggressive or perverted behaviour by amphetamine is hardly conducive to a satisfactory sex life.

Cannabis

Cannabis, in various forms, has been used as a drug for centuries, parti- cularly in the Middle East, Asia and North Africa. More recently it has been introduced into Europe and North America where it has become something of a cult. It usually consists of the flowering tops of the Indian hemp plant (*Cannabis sativa*), which grows widely throughout the warmer regions of Asia, Africa and America. It is illegal in Britain to possess any part of the plant except the seeds, which are readily available as bird seed in pet shops. Consequently, a considerable num- ber of illicit cannabis plants appear on window sills and in greenhouses in spite of the legal penalties. Cannabis appears in a range of prepara- tions under a variety of different names. The dried resinous exudate obtained from the flowering tops is often called hashish or, in the Far East, *charas*. The dried leaves and flowers, used like tea or tobacco, are called *bhang*, or, in the US, marijuana. Other contemporary street names such as pot, hash, grass, hemp or dope have also proliferated in North America and Europe.

The earliest reports of the use of cannabis appear in Chinese writings. The surgeon Hua-t'o apparently used it as a sedative and analgesic, although he makes no mention of its euphoric or aphrodisiac properties. It has been used for centuries in many Islamic countries in the way that alcohol is used in western countries, that is, as a social stimulant and panacea. Cannabis evidently shares with alcohol the ability to depress the inhibitory processes in the brain (disinhibition) at low dose. Higher doses apparently produce a soporific or even hallucinatory state in the subject.

In the Arabian Nights there is the tale of the *kazi* and the *bhang* eaters in which the recipe for *bhang* is given:

'tis composed of|hemp leaflets, where to they add aromatic roots and somewhat of sugar, then they cook it and prepare a kind of confection which they eat, but whoso eateth it (and especially an he eat more than enough) talketh of matters which reason may in no wise represent.

This production of nonsensical chatter, together with uncontrolled and inexplicable hilarity, is one of the peculiar features of cannabis intoxication.[4]

One of the fullest, and perhaps even the best, accounts of the effects of hashish intoxication is in the writings of Dr Jacques-Joseph Moreau de Tours, in particular in his book *Du hachisch et de l'alienation mentales. Etudes psychologiques*. This was originally published in Paris in 1845 but has since been translated into English and reissued.[5] The French writer Theophile Gautier, in his essay 'Le Club des Hachischins', mentions a certain Doctor X who had travelled widely in the Orient. It seems likely that this character was, in fact, Dr Moreau. Le Club des Hachischins existed as a group of writers, painters, sculptors and other artists who met regularly at the Hotel Lauzun on the Isle St. Louis in Paris with the primary object of indulging in cannabis. The members included the writers de Balzac, Baudelaire and Dumas, all of whom took hashish, usually in the form of an electuary, in order to experience its intoxicating and psychedelic effects. Moreau, as a medical man, was on hand to keep an account of the effects of the drug both on himself

Figure 7.2: An Example of a Privately Sponsored Card Warning of the Dire Hazards of Marijuana

(Sample--Warning card to be placed in R. R. Trains, Buses, Street Cars, etc.)

Beware! **Young and Old—People in All Walks of Life!**

This **may be handed you**

by the <u>friendly stranger</u>. It contains the Killer Drug "Marihuana"-- a powerful narcotic in which lurks
                        Murder!   Insanity!   Death!

WARNING!

Dope peddlers are shrewd! They may put some of this drug in the 🫖 or in the ᶜᵒᶜᵏᵗᵃⁱˡ or in the tobacco cigarette.

WRITE FOR DETAILED INFORMATION, ENCLOSING 12 CENTS IN POSTAGE — MAILING COST

Address: THE INTER-STATE NARCOTIC ASSOCIATION
(Incorporated not for profit)
53 W Jackson Blvd.   Chicago, Illinois, U. S. A.

and on the fellow members of the club.

In his notes, Moreau describes the psychological effects of hashish as a sequence of phenomena: first, a feeling of happiness; second, excitement, dissociation of ideas; third, errors of time and space; fourth, development of the sense of hearing; the influence of music; fifth, fixed ideas (delusions); sixth, damage to the emotions; seventh, irresistible impulses; eighth, illusions and hallucinations.

Apart from their accuracy, Moreau's observations are all the more remarkable in the light of the primitive state of psychological and behavioural medicine at that time. His main thesis was that the effects of hashish resembled, in many ways, several of the behavioural changes associated with mental illness. He considered that hashish might be useful in the treatment of some forms of depression, as well as in cases of manic excitement. (A similar principle was advanced to support the use of LSD in treating schizophrenics over a hundred years later.)

In all the surviving accounts of the activities of the Hashish Club, however, no aphrodisiac effects are ever mentioned. It is possible that, in the moral climate of the time, Moreau felt unable to mention such matters, but in view of his highly objective attitude towards these experiments, this seems very unlikely.

Cannabis contains a large group of chemicals known as cannabinols, of which $1\text{-}\Delta^9$-tetrahydrocannabinol (THC) now appears to be the active constituent:

Cannabinol
(inactive)

Tetrahydrocannabinol
(active)

This drug has been synthesized, together with a large number of close structural analogues, some of which are more potent than the parent compound. Scientific interest in the cannabinols has arisen as a result of the sudden spread of cannabis usage from the poor black ghettoes of the larger cities in the US to the middle-class white youths of both America and Europe. It has been estimated that in the US by 1972, 16 per cent of the adult and 14 per cent of the juvenile population had used cannabis at least once in one form or another. Part of the vicarious interest in the drug has no doubt come about purely because it is illegal to possess it or grow it, and to use it constitutes a challenge to the accepted rules of society which, in its wisdom, accepts the use of a physically addictive drug (alcohol) and a cancer-producing and habit-forming plant (tobacco). In fact, the recreational use of cannabis has not been prevented by severe legal penalties, and from a scientific point of view it is highly questionable whether its continued suppression by the law is justified on the basis of its pharmacological and toxicological properties. On the other hand, although the occasional use of cannabis may be relatively safe, its continued use to excess could well be dangerous as with alcohol or tobacco. The cost to governments of the damage done to society by the last two drugs is generally believed to outweigh the income derived from taxing them, and to add yet another drug to social and fiscal respectability is perhaps morally unjustifiable.

The behavioural effects of cannabis have been notoriously difficult to measure by any acceptable scientific method. Many of its supposed actions have been communicated by word of mouth and hearsay, and it is not really surprising that it has gained a reputation as an aphrodisiac. The Indian Hemp Drugs Commission of 1894 reported that hemp drugs: 'have no aphrodisiac power whatsoever; and as a matter of fact, they are used by ascetics in this country [India] with the ostensible objects of destroying sexual appetite'. More recent clinical investigations have produced results that support this view. In one study,[6] groups of 20 heterosexual males aged between 18 and 28 years were matched with similar groups of chronic cannabis users. None of the subjects was taking any other drugs apart from cannabis in the form of cigarettes or reefers. Sperm counts and blood plasma levels of testosterone were measured at intervals. Those subjects smoking five to nine reefers a week showed a 40 per cent fall in plasma testosterone levels, and those smoking more than ten a week showed an even greater reduction. Abstention from cannabis and treatment with gonadotrophin drugs were able to restore the testosterone levels to normal. The heavier smokers also showed significantly reduced sperm counts, and two of these subjects actually reported that they were impotent. One of the pair gave up smoking in consequence and was subsequently able to report to the investigators that he no longer had any difficulty in achieving an erection.

The above results obviously suggest that cannabis is very unlikely to increase either sexual potency or ability. Many of the comments made by cannabis users to scientific investigators at interview have included such statements as, 'Although I have erotic thoughts and dreams, I don't actually feel like doing anything.'

In contrast to alcohol, cannabis seems to reduce aggression and produce a feeling of well-being (although a small proportion of first-time users apparently suffer from feelings of panic and anxiety).

On balance, cannabis seems more likely to produce an aphrodisiac effect in the female sex rather than the male, although quantitative assessment of this is at present impossible.

The aim of a cannabis smoker is usually described as 'getting a high', although this is not the same type of high as that associated with amphetamine, for example. It requires an appropriate environment, like-minded company of either sex, and a certain degree of experience in smoking a reefer or 'joint'. Different sources or varieties of cannabis are credited with particular properties and the rate of burning of the reefer and the proportions of air and smoke drawn into the lungs are all important variables in the ritual. The subjective effects of the drug vary widely between individuals, and the concentration of THC in the reefer is also an unknown and probably highly variable quantity. Below is a list of the most frequent symptoms reported by a group of 111 subjects given either cannabis or a placebo.[4]

Marijuana (74 subjects)	*Placebo (37 subjects)*
Noticed passing of time more	Body relaxed
Time going slower	Unsteady
Speech sounds slower	Sluggish
Thoughts moving faster	Felt high
Imagination more lively	Sleepy
Heart beating faster	More relaxed
Throat drier	Felt more at peace with the world
Hungrier than usual	Losing sense of time
Thinking clearer	Thoughts moved slower
Easier to concentrate	Noticing passing of time less
Felt more free	
Felt more serious	
Noticing things more	
Colours brighter	
Arms and legs more sensitive	
Disliked answering these questions	

The level of expectation and suggestion among the subjects is clearly revealed by these subjective self-assessments, and the effectiveness of the placebo is well demonstrated. The section on the use of placebos in Chapter 6 discussed this phenomenon in more detail.

What is perhaps a little disappointing (to those who might have expected it) is the lack of any mention of aphrodisiac feelings, however vague. Possibly none of the subjects in the above study believed in, or was expecting, an aphrodisiac effect.

Sexual Behaviour of Cannabis Users

A report in 1971[7] described severe adverse psychological effects in a group of 38 cannabis users who reported for psychiatric treatment. In many cases, the symptoms disappeared when the patients gave up using the drug. In a group of 18 unmarried women aged between 13 and 22 years, one was psychotic and the remainder showed promiscuous activity; seven became pregnant and four contracted venereal disease. It should be mentioned also that most of these subjects exhibited confusion, apathy, depression (sometimes to the extent of attempted suicide), feelings of isolation and generally inappropriate behavioural responses. This does not mean that cannabis has all these effects as well

Table 7.1: Correlation between Marijuana Smoking and Premarital Sex

Frequency of marijuana smoking	Number of Subjects	Percentage Indulging in Premarital Sex	
		Yes	No
Over three times a week	89	85	15
Two times a week to 3 times a month	155	74	26
Less than three times a month	133	65	35
Never in last six months	167	37	63

as promoting promiscuity; it is far more likely that other psychological disturbances, revealed by the disinhibiting effect of the cannabis, led to this type of behaviour. The number of subjects (18) was small, they were all in a similar adolescent age group when sexual awareness might be expected to be at a peak, and they were all originally selected because of their psychiatric problems rather than for the fact that they smoked cannabis.

A similar pattern of behaviour is often found in other groups of drug users, notably amphetamine or narcotic addicts, who have severe personality problems which they try to solve by escaping into a drug habit. A questionnaire study carried out on an American college campus, involving a more 'normal' group of subjects, revealed a strong correlation between the incidence of marijuana smoking and premarital sexual relations among the college students.[8] The relevant details are shown in Table 7.1.

From the results in Table 7.1 it is apparent that the more frequently a student smokes marijuana, the more likely he or she is to indulge in sexual activity. A similar correlation was found to exist between the frequency of marijuana smoking and the number of sexual partners. In the face of these statistics it is tempting to jump to the conclusion that marijuana smoking induces sexual promiscuity, although there is no evidence of any causal relationship between the two activities. It is far more probable that the sort of people who are prepared to indulge in an illegal activity such as marijuana smoking are also liable to have a similarly broad-minded attitude towards sexual *mores*, and be unlikely to adhere to traditional sexual codes. In other words, the more sexually adventurous members of society are more likely to be marijuana smokers.

Cocaine

The Incas of South America worshipped the sun as the father of the gods and the coca plant as a divine gift from the mother of the gods, Mama Coca. This story of the divine origin of the plant may merely have been a ploy to ensure that it was reserved for priestly use, but even before the Inca civilization had appeared, the Moche people had discovered and exploited the stimulant properties of coca leaves. Moche art and sculpture had a very conspicuous sexual theme, with bestiality, fellatio, homosexuality and so on graphically represented on articles of pottery and sculpture. To the later European explorers the appearance of such obviously perverse and degenerate sexual behaviour in a culture which also used the coca leaf for its pleasurable effects suggested that the two practices were related. In other words, coca use would lead to sexual excess.

The second Council of Lima in the middle of the sixteenth century attempted to restrict the use of coca, describing it as 'a useless object liable to promote the practices and superstitions of the Indians'. The Conquistadores, however, who were exploiting the local Indian labourers in mines and plantations, did not hesitate to pay their workers in coca leaf. It was cheap, it increased the workers' stamina and it made them more capable of standing up to the rigours of forced labour and a poor diet. The isolation of cocaine from the coca leaf and its introduction into medical use towards the end of the last century encouraged the governments of Bolivia, Peru and Colombia to make the coca industry a state monopoly so that its worldwide sale and distribution would be under central control, to the immediate benefit and profit of the governments concerned.

The History of Cocaine

The first European to discover the shrub *Erythoxylon coca*, from which cocaine is derived, was Francesco Pizarro who set out to explore the interior of Peru in 1533. The local natives spent much of their time chewing the leaves in a mixture with lime or vegetable ashes. It was a popular practice in the regions closest to the Andes, one of the effects of cocaine being to relieve some of the unpleasant symptoms associated with breathing the oxygen-poor air of high altitudes. The Spanish Conquistadores were greatly impressed by the effects of chewing coca leaf and introduced it into Europe. The reason that it did not immediately become popular is probably due to the fact that tobacco was discovered and introduced at about the same time, and it was, perhaps fortunately,

Figure 7.3: An Advertisement from the *St Louis Journal of Medicine* for a Tonic Wine Containing Cocaine. The 500 ml bottle sold at 5 francs and was only one of the many cocaine-containing nostrums on sale at the turn of the century. (Courtesy of Dr Robert Byck, Yale School of Medicine)

"Mariani Bottle" showing Shape and Label.

"Mariani Bottle" showing Outside Wrapper.

We are justified in saying : Never has anything been so highly recommended and every trial proves its excellence.

Size of Regular Bottle, half litre (about 17 ounces).

Never sold in bulk– to guard against substitution.

VIN MARIANI

Nourishes - Fortifies
Refreshes
Aids Digestion - Strengthens the System.

Unequaled as a tonic-stimulant for fatigued or overworked Body and Brain.

Prevents Malaria, Influenza and Wasting Diseases.

We cannot aim to gain support for our preparation through cheapness ; we give a uniform, effective and honest article, and respectfully ask personal testing of **Vin Mariani** strictly on its own merits. Thus the medical profession can judge whether **Vin Mariani** is deserving of the unequaled reputation it has earned throughout the world during more than 30 years.

Inferior, so-called Coca preparations (variable solutions of Cocaine and cheap wines), which have been proven worthless, even harmful in effect, bring into discredit and destroy confidence in a valuable drug.

We therefore particularly caution to specify always " VIN MARIANI," thus we can guarantee invariable satisfaction to physician and patient.

tobacco that became the popular recreational drug in Europe.

The Spaniard, Nicholas Monardes, wrote of the marvels of coca in his book *Joyfull Newes out of the Newe Founde Worlde*, published at the end of the sixteenth century:

> I was desirous to see that hearbe so celebrated of the Indians, so many yeres past, which they do call the Coca, which they doe sow and till with muche care and diligence because they use it for their pleasures, which we will speake of.
>
> The use of it amongst the Indians is a thing general for many things, for when they travell by the waye, for neede and for their content when they are in their houses they use it in this forme. They take Cockles or Oysters, in theyr shelles and burne them and grinde them, after they are burned they remaine like Lime, very small ground; then they take the leaves of the Coca, and chewe them in theyr Mouthes, and as they chewe it, they mingle it with some of the pouder made of shelles in such sorte, that they make it like a paste, taking less of the Pouder than of the herbe, and of this Paste they make certeyne small bawles round, and lay them to drie, & when they will use them, they take a little Ball in their mouth and chewe it rowling it from one place to another. . .
>
> For the use of these little Balles taketh the hunger and thirst from them: & they say that they receive substaunce thereby, as though they did eat meat. At other times they use them for their pleasure . . . When they will make themselves drunke, and be out of judgement, they mingle with the Coca the leaves of the Tobaco, and chewe them altogether, and goe as it were out of their wittes, or as if they were drunke, which is a thing that dooth give them great contentment, to be in that sort.

It was not until 1859, when the Italian physician Paola Montegazza, who had travelled widely among the Peruvian Indians, published a report on the potential medical uses of coca, that European doctors were alerted to the properties of this new wonder drug. Shortly afterwards, the Austrian chemist Alfred Niemann succeeded in purifying an active alkaloid from coca leaves which he named cocaine. There followed a brief honeymoon between the medical profession and Mama Coca before the dangerous and addictive properties of the drug were recognized and its use was terminated.

Freud and Cocaine

Cocaine has had numerous proponents, from Sigmund Freud to Errol Flynn, but there is no doubt that the brief popularity of cocaine as a universal panacea at the end of the last century was largely due to the eloquent advocacy of the Viennese physician. Freud had read about the properties of cocaine in 1884 and, in the same year, ordered a small quantity from the Merck company at Darmstadt. He experimented upon himself and found that the drug produced euphoria and loss of appetite without diminishing his energy or vigour. Freud missed the principal clinical application of cocaine as a local anaesthetic in opthalmic medicine (this was taken up and exploited by his colleague Carl Koller), and instead introduced it to his friend Ernst von Fleischl-Marxow who had become addicted to morphine during a prolonged course of treatment for cancer. The aim was to facilitate the gradual withdrawal of the morphine, and the results were initially promising, but the unfortunate Fleischl rapidly became addicted to the cocaine and was soon taking 1 g per day, one hundred times the dose which Freud had originally tried himself.

Freud had committed himself wholeheartedly to the promulgation of cocaine as a universal panacea, and his initial successes with cocaine led him to write:

> If it goes well I will write an essay on it and I expect it will win its place in therapeutics, by the side of morphine and superior to it. I have other hopes and intentions about it. I take very small doses of it regularly against depression and against indigestion, and with the most brilliant success.

Freud also gave some of the drug to his fiancée Martha 'to make her strong and give her cheeks a red colour', but, despite writing the promised essay 'In Praise of Coca', his predictions remained unfulfilled as the addictive potential of the drug was realized. By 1886 Freud was under attack for his unreserved recommendation of cocaine, Erlenmeyer even referring to the drug as the 'third scourge of mankind'.

In the essay in question, Freud had suggested the following therapeutic applications for cocaine:

(1) to increase the physical capacity of the body for a given short period of time and to hold strength in reserve for wartime;

(2) in digestive disorders of the stomach;

(3) in cachexia (extreme states of general ill-health);

(4) to counteract the morphine withdrawal syndrome;
(5) in asthma;
(6) as a sexual stimulant or aphrodisiac;
(7) as a local anaesthetic.

The episode with cocaine has always remained as something of a black mark against the career of an otherwise brilliant innovator, and Freud was eventually forced to withdraw his support for the drug. The consequences of the increased use of cocaine by Europeans and Americans is reflected in the figures for the annual export of coca leaves from Peru and Java (where it was introduced) over this period:

	Peru		*Java*
	kg		*kg*
1877	8 000	1904	26 000
1906	2 800 000	1911	740 000
1920	453 000	1920	1 700 000

The current traffic in cocaine from South America is mostly illicit and it is therefore very difficult to make accurate estimates of the rate of production. There is still a tendency to assume that, because cocaine can be taken orally or nasally, it is somehow less dangerous than the injectable narcotic analgesics. On the contrary, there is ample evidence that cocaine is a dangerous drug with strong addiction potential.

The Pharmacology of Cocaine

Cocaine is benzoylmethylecgonine, an ester of benzoic acid and a nitrogenous base chemically related to atropine. It has a rather complicated and unique pharmacology; no other drug combines the local anaesthetic and central stimulant properties of cocaine. Its principal action on the central nervous system is to prevent the re-uptake of noradrenaline into the neurones from which it has been released as a transmitter. Since these noradrenergic neurones have mainly excitatory and stimulant actions, their function is potentiated by cocaine. This is the basis for the increased mental activity, restlessness and euphoria that cocaine produces. In addition it increases the respiratory rate, the heart rate and the blood pressure. In many ways the effects of cocaine are very similar to those of amphetamine, although their precise mechanisms of action are slightly different.

The toxic effects of cocaine include muscular tremors and convulsive movements, nausea, vomiting, raised body temperature, and hallucinations, which can be visual (objects appearing to move or pulsate), tactile (feelings of small animals crawling under the skin), olfactory (strange and inappropriate odours perceived), or auditory (hearing voices). In some ways the hallucinations resemble the aura associated with impending migraine attacks and in others they are more like the delirium tremens that occurs during alcohol withdrawal. Death from cocaine poisoning is usually due to acute heart failure or respiratory arrest. There is no completely reliable treatment for acute poisoning.

Cocaine as an Aphrodisiac

Cocaine and sex have become inextricably linked ever since the time of Freud's encouraging essay in 1884. Several less than objective accounts of the sexual athletics that have been made possible by the use of cocaine, notably the *Diary of a Drug Fiend* by Aleister Crowley, have served to further this association. At the end of the last century cocaine was being included in a large number of tonic-stimulant drinks. Mariani wine was a very popular nostrum which was claimed to nourish, fortify, refresh and strengthen the system. Other preparations which flourished at the time included Koca-Nola, Cafe-Cola, Kos-Cola, Celery Cola, and Dr Don's Kola, but the most famous tonic drink containing cocaine is still with us, at least in name. 'Coca Cola', introduced by Asa G. Chandler, meant exactly what it implied; this was a drink made with extracts of coca leaves and cola nuts. The cocaine has long since been removed from the recipe, but the term 'Coke' is still used interchangeably for the drink and the purified drug.

The end of the popular cocaine era came with the introduction of the Harrison (Narcotics) Tax Act of 1914, which required any persons authorized to handle or manufacture drugs to register with the authorities, pay a fee, and keep accurate records of all the 'narcotics', including cocaine, in their possession. From this time on, cocaine went underground. The extract of cola (or kola) nut is still included in the modern versions of 'Coke' and 'Pepsi', along with various artificial colourings, flavourings and sweeteners, but the stimulant component is now provided by caffeine.

Cocaine can now only be obtained by illicit means, but the huge trade in the drug has recently resulted in a shortage of leaves for legitimate (and presumably less profitable) processing. The contemporary street names for cocaine are imbued with sexual innuendo: 'La Dama Blanca', 'Jam', 'Blow', 'Lady' and 'The Pimp's Drug', for example, and it has been described by regular users as the champagne of sexual drugs. To what extent is this remarkable reputation justified? It is probably best to consider first what cocaine can do, then examine the expectations of the users, and finally assess the extent to which these are fulfilled.

Used as a local anaesthetic and applied to the genital area, cocaine initially produces a not unpleasant tingling sensation until full anaesthesia is achieved. The deadening of sensation in the penis will assist in delaying ejaculation and enables the user to indulge in sexual marathons without losing his erection. The local action of cocaine to constrict blood vessels could also be of assistance in maintaining an erection, but it would tend to work against the normal nervous processes which are involved in achieving an erection in the first place (see page 146). Applied to the female genitalia, the loss of sensation tends to reduce sexual pleasure, but at the same time it will block any sensation of pain that might be caused during intercourse. The difficulty with this use of cocaine is that users need to employ increasingly forceful forms of sexual stimulation in order to feel anything, and physical injury to the genital regions is not uncommon.

The stimulant actions of cocaine will induce feelings of heightened mental and physical power. It is clear that cocaine will increase alertness and physical stamina while at the same time giving a sense of euphoria, and although this can obviously be an advantage in a sexual situation, there is no evidence that cocaine actually increases physical strength.

The levels of sexual expectation among cocaine users are high: of 53 per cent of males who used cocaine for a specific purpose, 42 per cent used it to enhance their performance at sports or sex. With females the figure was only 20 per cent. This perhaps reflects the fact that males are more likely to achieve some benefit from the use of cocaine, a view supported by the results of a survey of cocaine use by masseuses in California.[9] These masseuses provide sexual extras for their clients, many of whom use cocaine to enhance their performance. When interviewed, the girls tended to agree that cocaine did little for them sexually and that they only used it to help them keep awake. Their drugs of preference were marijuana and Mandrax.

Cocaine as an Anaphrodisiac

With cocaine, as with all the illicit drugs, there is the basic problem of assessing the dose of pure drug that people are receiving. The crudest form of cocaine available is coca paste (known as 'brute'), which is an extract of coca leaves and contains numerous contaminants including benzoic acid, ecgonine, methanol, kerosene, alkaline ash and sulphuric acid. It is normally mixed with tobacco and smoked. Even the supposedly pure base will get 'cut' or diluted with other drugs or inert substances in order to increase the profit margin for the dealer. Another particular problem with cocaine is that, although the effective nasal dose for obtaining euphoria is about 10 mg, symptoms of toxicity can occur with doses of anything from 20 mg to 1 g, depending upon the sensitivity of the individual and his degree of tolerance developed from previous use of the drug.

Early reports which described the sexual enhancement obtainable from cocaine also mentioned that its excessive use could lead to impotence in males and sterility in females. This view encouraged the *American Journal of Pharmacy* in 1903 to report: 'One redeeming feature there is, the habit of cocaine seems to lessen both sexual desire and ability, so there is less danger of its transmission by heredity.'

There have been virtually no studies on the effects of cocaine on sexual performance and those that have been published rely on user reports. These reports tend to overestimate the positive effects of the drugs in question, so that when negative reactions appear with any frequency, it can be assumed that the drug is having significant detrimental effects. The undesirable acute effects reported by cocaine smokers include slurred speech, thirst, anorexia, difficulty in focusing the eyes, difficulty in concentrating and tremors. Chronic effects include, in addition, dry and chapped lips, black sputum, chest and back pains, palpitations, dysphoria, hallucinations and feelings of paranoia.

In a study of 32 cocaine smokers carried out by Siegel[10] and consisting of 23 males and 9 females, 20 of the males experienced sexual impotence although 4 of the females reported improved sexual performance. When the sexual partners of the subjects were interviewed, Siegel found that almost all the cocaine smokers were experiencing episodes of loss of interest in sex. The anaphrodisiac effect of cocaine seems therefore to be associated with the prolonged use of the drug. It has also been established that, once started, the cocaine habit is extremely difficult to give up. Bearing this in mind there is no reason to use cocaine for its local anaesthetic effects since safer drugs are available quite legitimately for this purpose. Also, using cocaine as a stimulant involves a severe risk to health and is unlikely to be successful in the long term. Lewin has neatly expressed this conclusion as follows:

'Those who believe they can enter the temple of happiness through this gate of pleasure purchase their momentary delights at the cost of body and soul. They speedily pass through the gate of happiness into the night of the abyss.'

Opium and the Narcotic Analgesics

Opium is one of the oldest drugs known to man; its ability to abolish pain and induce sleep made it an invaluable aid to the early physicians who had precious few drugs to which they could turn. The word narcotic is derived from the name of the Greek god of sleep, Narcos, and indicates the principal use to which opium or laudanum was put. Other historic uses of opium included the prevention of diarrhoea, and the treatment of snakebite, epilepsy, asthma, coughs and headache. It has even been used to 'cure' the common cold. We now know that opium, the dried exudate obtained from unripe poppy heads, contains at least 20 active constituents, but principally morphine. Heroin (diacetylmorphine) does not occur naturally and is synthesized from morphine; it was originally introduced as a safer and less addicting alternative to morphine! The third of the narcotic analgesics that is widely abused is methadone. This is a purely synthetic compound which produces a milder withdrawal syndrome than either morphine or heroin and is used to wean addicts off their habit.

Opium abuse originally consisted of smoking the resin in a pipe and inhaling the smoke (chasing the dragon) in order to produce a semi-hypnotic dreamlike state in which the pain and cares of the world dissolved away. The preparation of pure morphine and the invention of the hypodermic syringe, however, changed the whole pattern of abuse. The clinical use of opiates usually involves administering suppositories or giving subcutaneous injections for the relief of acute pain; morphine has little effect on the brain when taken orally. The addict, on the other hand, in order to achieve the maximum 'high' from his dose of narcotic, almost invariably injects it intravenously. This produces a 'rush', a short-lasting feeling of intense euphoria and heightened sensations which some addicts have compared to a sexual orgasm. The chronically addicted user of narcotics is not necessarily seeking relief from pain, but is principally concerned with avoiding the pain and discomfort arising from the withdrawal syndrome which occurs as the blood concentration of the drug falls. An addict who can obtain a regular supply of the drug should therefore be able to lead a fairly normal life, allowing for the fact that tolerance to the drug will develop and progressively larger doses will be required to obtain the same effect.

The violent crime and prostitution associated with drug addicts are due to their inability to pay the high black market prices for the drugs from the dealer, and the consequent need to obtain money by any means possible. Respectable or wealthy members of society who can afford to buy their supplies on the black market, or members of the medical profession who can obtain the drugs through legitimate sources, have no need to turn to crime to finance their addiction. Besides the well-known figures of De Quincey and Coleridge, both Sir John Erskine and William Wilberforce were opium addicts who lived well into their seventies. Smoking opium is clearly much safer than injecting opiates with the attendant risks of septicaemia or hepatitis from dirty needles.

Opium as an Aphrodisiac

Opium was used and abused in Asia for centuries before it was introduced into Europe. One of the earliest accounts of its properties was written by the Portuguese physician, Garcia d'Orta (1501–1568), who, in 1563, published a series of colloquies on simple drugs in India.[11] His book was arranged as a series of imaginary conversations in which the properties of the drugs are explained to an interviewer.

The disinhibiting effects of 'bangue' (hashish) are accurately described, as are the hallucinatory effects of *datura* (hyoscine). On the subject of opium, or *amfium*, to give it its old name, the inherent paradox in its popularity as an aphrodisiac and its established anaphrodisiac properties is explained in some detail. The questioner in the following interview is called Ruano:

Ruano. And they do not take it for lustful purpose, so they tell me, as this is contrary to medical findings and all reason to think it is efficacious for the work of Venus.

d'Orta. What you say is true. Indeed it is not efficacious for this purpose, but actually somewhat harmful . . . and all the learned physicians of our rank tell me that it caused impotence in men and they soon became incapable for perform the act of Venus.

Ruano. But so many people use this for fleshly lust, they cannot all be deceived.

d'Orta. I will tell you what is useful for, if you will allow me, for this is not a very proper subject, especially when we discuss it in Portuguese.

Ruano. I believe that things are not indecent until they are mentioned by dirty-minded people and for indecent reasons.

d'Orta. The imaginative virtue plays a great part in carnal lust, and it is stronger than the expulsive virtue that follows it. Hence the act of Venus is completed all the more rapidly. And because the imaginative virtue is so powerful it dominates the expulsive one forcing into the testicles the genital seed, and the greater the imagination the more rapid the emission of the seed. And as those who have taken *Amfium* have lost control, they actually complete this venereal act much later. And because many females do not give away the seed that quickly, and when the man is also slow she completes the act of Venus much later, the act of conception ends exactly at the same time for both of them, and taking *Amfium* is here a help. It also assists to complete the venereal act more slowly; the *Amfium* also opens the ways by which the genital seed comes from the brain by reason of its coldness, and thus brings about its production in both simultaneously. I know that you understand this very well, but if you write it down in plain language it does seem rather an improper thing to do.

Ruano. So, they do have some excuse, though not a very honest one.

Apart from the mistaken belief in the source of the seminal fluid which, though it might seem rather bizarre today, is quite understandable bearing in mind the limited knowledge of human physiology at that time, it is difficult to improve on d'Orta's account of a drug-induced delay in ejaculation as a means of achieving simultaneous orgasm. The dose of opium consumed would have to be carefully judged, since its sedative effects would eventually render the subject effectively impotent, as indeed d'Orta noted.

Writing at about the same time, Cristobal Acosta (*c.* 1525-1594) also described this specific effect of opium on coition, and even quoted the work of d'Orta. He made some additional observations upon the widespread use and addictive properties of opium which are very relevant today:

The use of opium is widespread in those regions [East Indies], for by sleeping or by semi-intoxicating themselves in sleep, they can for a time free themselves from all anxieties and weariness; it is also used for venereal purposes. Although contrary to reason, it is used so widely for this purpose, indeed to such extent that it is the most common and ordinary remedy used by the vile children of Venus. It is believed not only by our own medical practitioners but also by other physicians, Arabs, Parsees, Turks, Coracons, Sundas, Malaysians,

Chinese and Malabars, together with all the Camarins, Decanins, Brahmins and other physicians, that the use of opium (by reason of its stupefying and narcotic quality) causes impotence in addicts, and this is confirmed by our reasoning and experience. But the imagination of the common folk is so powerful that it can transform impotence into potency, and they commonly employ it to increase their lascivious carnal pleasures, and what is even worse once the habit has been established, they cannot give up their pleasure and addiction without great risk to their lives.

More extensive extracts from these early writings on the psychopharmacology of drugs of abuse have been quoted by Guerra.[12] It is extremely interesting that the hazards inherent in the use of opium should have been recognized for over 400 years and that modern scientific investigations are serving only to confirm and reinforce these early empirical observations.

Contemporary narcotic addicts usually report a decreased appetite for, and interest in, food, drink and sex. In fact, they generally seem to be existing only for their next fix. The prostitute who is addicted is certainly not in the profession in order to satisfy a drug-induced appetite for sex. It seems that heroin and the other opiates are, in fact, providing a substitute for sexual pleasure rather than an increase in appetite or ability, and narcotics are basically anaphrodisiac, reducing the pleasure gained from sex.

Opiates as Anaphrodisiacs

Modern investigations into the behavioural effects of opiates have normally involved interviews with addicts who have a natural tendency to be mendacious. The results of surveys of their sexual activities should therefore be regarded with some degree of caution. However, they are more likely to overestimate their sexual prowess than be over-modest, so the studies that report a high incidence of impotence and diminished libido are more than likely to represent the true state of affairs.

In one study, published in 1972,[13] 61 per cent of heroin addicts reported impaired libido and 39 per cent admitted impotence. An independent study produced figures of 72 per cent and 25 per cent, respectively. Their subjects' specific problems included delayed ejaculation (70 per cent of addicts) as well as premature ejaculation. The latter may be a symptom of withdrawal from the drug rather than a direct effect. A specific example of the sexual problems faced by an addict have been described by Parr,[14] who interviewed an addict and his wife:

> After his large morning fix of heroin, which used most of his day's supply, he was at first impotent. Then, during the day, he developed erections, but could not ejaculate. Later, libido increased and normal responsiveness developed until in the early evening his potency was excellent with optimal ejaculatory time. During the evening his sexual reflexes became quicker and *ejaculatio praecox* was the rule. During the night, the run-down of drugs in his system was associated with spontaneous ejaculations. The same cycle repeated itself every 24 hours. The couple programmed their love life accordingly.

Reports with methadone are essentially similar to those from addicts taking heroin. The incidence of decreased libido has ranged from 6 to 38 per cent, and failure of erection has been reported in up to 50 per cent of users. There is no reason to believe from this large amount of evidence obtained over a long period of time that the opiates are anything other than anaphrodisiac in their effects.

Nicotine

Nicotine is one of the few naturally occurring liquid alkaloids, and is the major active constituent present in tobacco leaves. Although it was not isolated until 1828, and its pharmacological properties were not studied until 1890, the supposed medicinal value of tobacco, either smoked or decocted, had been recognized virtually since its introduction into Europe by Jean Nicot, the Ambassador of the King of France to Portugal from 1559 to 1561.

The pharmacological effects of nicotine on the autonomic nervous system have been thoroughly investigated, and its basic mode of action is well understood. It acts to mimic the transmitter acetylcholine at both the sympathetic and parasympathetic ganglia, and at the junction between the motor nerves and the voluntary muscles. However, in the continued presence of the drug, nerve transmission is eventually blocked as the acetylcholine receptors become desensitized to further stimulation. Because of the variation in the time course of the onset of these two phases of action, and the varying sensitivity to the action of nicotine between specific ganglia, the overall effects observed can be complex and inconsistent.

The effects of nicotine that can be detected in a non-smoking subject who inhales a cigarette are, principally, an increase in heart rate, a rise in blood pressure, and a peripheral vasoconstriction that causes the extremities of the limbs to feel cold to the touch. Over a longer period of time a marked antidiuretic effect can be observed, due to stimulation of the release of antidiuretic hormone from the pituitary. Nicotine also appears to stimulate the emetic trigger zone, producing feelings of nausea. Experienced or heavy smokers are tolerant to all these effects of nicotine. As well as producing tolerance, nicotine also produces a very strong psychological dependence, and giving up smoking can be an extremely difficult process.

Nicotine has been used in the treatment of a bewildering range of disorders, from chronic itching and worms, to psoriasis, hiccups, muscle spasms, and even epilepsy. It has no current clinical use, mainly because of its high toxicity: on an equivalent weight basis it is more poisonous than cyanide. There have been numerous causes of nicotine poisoning as a result of the absorption of the drug through the skin from insecticidal preparations, and the acute lethal dose for a healthy adult can be as little as 60 mg, equivalent to about ten cigarettes. Obviously not all the nicotine in a cigarette is absorbed during smoking. The accidental swallowing of tobacco is rarely fatal, and its strong emetic action ensures that most of the ingested dose is lost by vomiting.

There is no evidence that nicotine possesses any aphrodisiac effects, although the major tobacco companies, through the medium of their extensive advertising campaigns, have attempted to persuade the public that the smoking habit is sexually attractive and socially acceptable. The recent stringent limits on cigarette advertising introduced by the British Government have tended to reduce this rather questionable method of marketing a potentially hazardous product. These restrictions do not at present extend to the advertising of pipe tobaccos, and we are still being entertained by the unlikely spectacle of women smiling ecstatically as pipe smoke is blown across their faces, or being slavishly attracted to the man with the right aroma.

The social etiquette of smoking has a long and interesting history. Until the First World War it was a strictly male indulgence, but the emancipation and enfranchisement of women led quickly to their freedom to smoke in public (they had probably been smoking in private for years). Initially, the type of girl who smoked in public was considered, at best, to be daring and, at worst, to be of loose morals. This association of dubious sexual *mores* with smoking in public gradually gave way to a fashionable acceptability that enabled smoking to become equated with social sophistication. Under-age smoking is still attractive to some by virtue of its illicit nature, and the use of tobacco by young people of both sexes presumably reflects an attempt to achieve a supposedly mature sophistication by the confident handling of a cigarette.

The sex equality of smoking habits does not yet extend to the woman who smokes a cigar or pipe. Women of this type are still viewed with great suspicion as regards the direction of their sexual interests. The pipe and the cigar have purely male connotations, and no tobacco company in the foreseeable future is ever likely to portray a man having smoke blown into his face by a pipe-smoking woman, or being slavishly attracted to her rich aroma. Our social conditioning has been very much determined by exposure to the stereotyped images projected by the advertisers.

Nicotine as an Anaphrodisiac

That smoking, as a social activity, can render a person more attractive or interesting to the opposite sex is open to debate, but there is no doubt that nicotine as a drug can act to cause sexual dysfunction in the male. One of the more dangerous side-effects from smoking is the reduction in blood flow to the extremities, that is, the fingers and toes. In severe cases the limbs become gangrenous and require amputation. This effect of nicotine can also be observed on the blood flow through the main artery to the penis, so making an erection more difficult to achieve. A recent report of two cases of young male smokers from Sweden illustrates this particular problem.[15] The young man, aged 20 and 27 years, were otherwise healthy but had impaired erectile ability. Measurement of their blood pressure and peripheral blood flow rate indicated a significant reduction in the blood pressure in the penile artery compared with that in the brachial artery to the arm. The two men gave up smoking, their penile blood flow returned to normal, and they regained a normal erectile capacity. Unfortunately, the authors gave no indication of how much or how long the two young men had been smoking. It would be interested to investigate the extent to which cigarette smoking might be contributing to the incidence of erectile impotence in the male population as a whole.

Nitrites

The first of the organic nitrite or nitrate salts to be introduced into clinical medicine was amyl nitrite, which was discovered by Dr T. Brunton in 1867 to relieve the immediate symptoms of *angina pectoris*. This acutely painful heart condition arises as a result of the constriction of the blood vessels supplying the heart. Inhalation of the highly volatile amyl nitrite produces an almost instantaneous relaxation of both arterial and venous smooth muscle. In addition to relieving the anginal pain, the nitrite also produces flushing of the face, due to the dilatation of the peripheral blood vessels. The side-effects include light-headedness (possibly caused by slight cerebral ischaemia as the blood is shunted into the large veins of the lower limbs), and this can be manifested as a sudden faint if the subject tries to stand up quickly after inhaling amyl nitrite. If the subject remains lying down, the dilatation of the cerebral blood vessels can produce an unpleasant throbbing headache. As might be expected, the drug produces sudden fluctuations in the blood pressure, and the electrocardiogram can show short-lasting changes in the electrical activity of the heart.

In normal medical use the advantages of amyl nitrite outweighed the lesser discomforts of its immediate side-effects. The only risk attached to its regular use was that it could affect the ability of the red-blood-cell haemoglobin to carry oxygen to the tissues. Amyl nitrite has now been superseded in clinical practice by longer-acting organic nitrates which provide a much more effective control of angina and congestive heart failure, so that acute emergencies occur very infrequently. Butyl nitrite was introduced about twenty years after amyl nitrite and has similar properties, but it never became widely used.

Amyl nitrite is a liquid which was dispensed in sealed glass ampoules enclosed in a cloth webbing. The ampoule could be easily crushed and the released vapour inhaled. The pop of the breaking ampoule gave amyl and butyl nitrite their generic street name of 'poppers'. The sales of amyl nitrite rose rapidly in the mid 1960s (it did not at that time require a prescription) and it was obvious that some recreational or hedonistic use had been found for it. But the volatile nitrites have no euphoric or psychedelic properties, and the first group of people to use the drug to any great extent were male homosexuals engaged in anal intercourse. A new aphrodisiac had been found.

Nitrites as Aphrodisiacs

The homosexual community had discovered that the smooth muscle relaxing properties of the nitrites applied not just to the blood vessels but also the anal sphincter. Inhaled by the passive partner at the moment of penetration, it would make intercourse easier and more comfortable. Dominant homosexuals also found that it could enhance their orgasm. The nitrites have subsequently been used by heterosexuals in the belief that they increase the blood flow into the penis and improve the erection, but there is no scientific evidence for this.

Nitrite preparations have become widely available to the gay community through advertisements placed in pornographic magazines and over-the-counter sales in sex shops. They can be sold openly as 'room odourizers' in bottles containing anything from 10 to 30 ml of liquid. In order for them to be sold legally, the label must state that the contents are not to be inhaled, although this is precisely how they are used. Commercial preparations include such brand names as 'Heart-On', 'Jac Aroma', 'Rush', 'Bullet', 'Locker Room', 'Aroma of Men Toilet Water' and so on. Over the period 1972–1977 it has been estimated that over 5 million bottles have been sold. This suggests that the nitrites are a very popular aphrodisiac indeed.

Users of butyl nitrite have reported that it prevents premature ejaculation, increases the sperm (seminal fluid volume?), improves the orgasm, and relaxes the anus.[16] A small minority of users reported that they did not enjoy using butyl nitrite during sex, particularly because of its various other side-effects, including light-headedness, dizziness, changes in their perception of reality, loss of identity, headache, nausea, and muscle weakness. The overall sensations produced by the drug will be a combination of not just its actual pharmacological effects, but also the level of expectation of the user, and the environment and circumstances under which it is taken.

There does not seem to be any convincing argument for the use of butyl nitrite by heterosexual couples, and the risks attached to its regular use far outweigh the marginal aphrodisiac properties it might possess. Anyone suffering from high blood pressure, anaemia, or any form of heart disease would be well advised to avoid it. Also, sufferers from migraine could well find that the drug is capable of inducing an attack.

Glyceryl Trinitrate

Glyceryl trinitrate is one of the drugs that have replaced amyl nitrite and it is available in the form of an aerosol, tablets for oral absorption, or a skin-absorbed ointment. There has been at least one case report involving glyceryl trinitrate in which a middle-aged male suffering from atherosclerotic impotence was being treated for depression.[16] The depression may well have been partly due to the impotence, but he had nevertheless been treated with the usual range of antidepressant and other psychotropic drugs, without any conspicuous success. However, at one stage the patient reported having chest pains, and the cardiologist whom he consulted prescribed glyceryl trinitrate. Two days after initiating this treatment the patient's waking erection returned (after a lapse of over two years), and he subsequently achieved successful intercourse and orgasm.

The beneficial action of the glyceryl trinitrate was presumably as a result of its vasodilatory properties, but its singular efficacy in this one case does not necessarily imply that it would be effective in all cases of organic impotence. There are several possible causes of impotence, but where the blood vessels have become narrowed, as in atherosclerosis, a vasodilator might be sufficient in itself to improve erectile efficiency. What is especially interesting in this particular case is that the trinitrate was not prescribed for the treatment of the impotence, and so any placebo effect, or suggestibility of the patient or the doctor, can be ruled out.

Nutmeg

Nutmeg, as used as a spice in the kitchen, is the nut of an East Indian tree, *Myristica fragrans*, which also yields the spice called mace. It has a particular reputation as an aphrodisiac among Chinese women and also in the Yemen. More usually it is employed to provide flavouring for sweets and puddings, and no suggestion of its venereal properties appears in the average western recipe book. However, in prison communities both here and in the United States, it is more widely used as a drug of abuse, particularly for the hallucinations it can produce at high doses. Many regular drug users, finding themselves imprisoned as a result of their habit, resort to nutmeg as a substitute for less readily available drugs. It is taken as a powder either by sniffing, like cocaine, or by suspending it in hot milk. The dose required to produce the symptoms of intoxication is usually in excess of two tablespoons of powder, equivalent to two grated nutmegs.

A detailed report of the effects of nutmeg has been published,[18] based on the account of an experienced user. The reported effects should be sufficient to discourage any use of nutmeg as a potential aphrodisiac; the initial side-effects were headache, severe nausea and dizziness lasting up to one hour, followed by hallucinations and other sensory disturbances which lasted up to two days. Other symptoms included dry mouth, lack of peristalsis (constipation), and an inability to pass urine. A profound and long-lasting hangover can also be expected. Experienced users have claimed that the psychedelic effects of nutmeg are similar to those of LSD.

Nutmeg contains a large number of identified organic compounds including between 5 and 15 per cent by weight of volatile oils and between 25 and 40 per cent of fixed oils. Of the volatile oil, 4 per cent is myristicin (5-allyl-1-methoxy-2,3-methylene dioxybenzene), and this seems to be the active principle. Some semi-synthetic derivatives of myristicin have been made, in particular MMDA (3-methoxy-4,5-methylene dioxyamphetamine) and MDA (3,4-methylene dioxyamphetamine). Both these drugs have found favour as psychedelics in the drug community of North America and are consequently featured in Schedule I of the uniform drugs legislation. MDA has been described as 'speed for lovers', and some of its effects are indeed similar to amphetamine but, unlike amphetamine, it tends to produce a relaxed feeling and a decreased desire for orgasm. The particular danger inherent in the use of MDA and MMDA is that the sensitivity of individual subjects to the toxic effects of the drugs varies widely, so that cases of accidental and fatal overdoses have occurred.

Nutmeg also has the doubtful reputation of being abortifacient, and it was probably with this in mind that a North American woman consumed 18.3 g of nutmeg in an attempt to induce menstruation. Her subsequent symptoms included spells of delirium alternating with stupor lasting 12 hours. She also reported feelings of fear and apprehension of impending death, and she required hospital treatment with tranquillizers. Periods of numbness of the extremities and dizziness lasted up to four days and the patient was only discharged from hospital a week after taking the nutmeg. Not for nothing is nutmeg known on the streets as the 'drug of last resort'.

The culinary use of nutmeg is unlikely to produce any untoward psychedelic experiences since the quantity used in a recipe is normally never more than 20 mg, or roughly one-thousandth of the dose taken by the woman in the previous report. At this dose no symptoms, either toxic or aphrodisiac, should be produced. The reputation of nutmeg, as with all other spices such as mace, cinnamon, allspice and so forth, lies in its ability to impart an exotic flavour to food or drink. In the right circumstances exotic dishes can have the most lubricious effects.

LSD and the Psychedelic Drugs

The term 'psychedelic' was coined by Osmond about 25 years ago to describe the properties of a group of drugs which had been found to alter sensory perceptions. Although many of them, notably LSD and mescaline, produce visual hallucinations, this is not necessarily true of the group as a whole. Since some of them can induce altered mental states which resemble certain forms of mental illness, it has been suggested that the term 'psychotomimetic' might be more appropriate. On balance, the term 'psychedelic' does seem to be the best way of describing these drugs. What they have in common is a high potential for abuse by the drug orientated community. There was a brief period when some of them were investigated clinically for the treatment of psychotic illness but, with the exception of ketamine, which is used as a general anaesthetic, they have now been generally discounted as having any therapeutic value.

Many of the psychedelics are natural in origin and have been used by priests or magicians in primitive societies for their capacity to induce trance-like states in which the priests can communicate with the gods. The Native American Church of North America has thus been able to obtain legal dispensation for the use of *peyote* (mescaline) as a sacrament in their religious services. Unlike most vegetable alkaloids, most of the psychedelics have fairly simple chemical structures related to amphetamine, and so their illicit laboratory synthesis has not presented too much of a problem to the entrepreneurs of the contemporary drug scene. A list of some of the drugs to be discussed, together with their properties, is given in Table 7.2.

The historic use of the naturally occurring psychedelics was principally for the mystical insights and experiences which they could induce, and their tribal use was often limited to a few chosen members of the tribe and then only at particular times in special ceremonies. In other words, they were never used recreationally as we would use alcohol or tobacco. Under these circumstances, no suggestion of any aphrodisiac effects was apparent until the hedonistic hippies of the 1960s took the opportunity of experimenting with these drugs as an alternative to heroin, cannabis, cocaine and the commonly available sedative-hypnotics. The reason for this awakening of interest can be traced to

Table 7.2: Psychedelic Drugs of Abuse

Drug Name	Source	Properties
LSD (lysergic acid diethylamide)	Synthetic	Blocks the actions of 5-HT (serotonin in the brain)
DMT (dimethyl tryptamine)	Synthetic, but also found in cohoba beans	As for LSD
Bufotenine	*Amanita muscaria*, toad skin, cohoba beans	As for LSD
Ololiuqui	Morning glory seeds (*Ipomoea*)	As for LSD, but also has amphetamine-like actions
Psilocybin	Several species of mushroom, notably *Psilocybe mexicana*	Blocks serotonin but is not as potent as LSD
Mescaline	Peyote (*Lophophora williamsii*), a type of desert cactus	Initially amphetamine-like; produces hallucinations
MDA (3,4-methylene dioxy-amphetamine)	Synthetic	As above
DOM (2,5-dimethoxyamphetamine) also known as STP	Synthetic	As above
Myristicin	Nutmeg ⎤	As above, but also have very unpleasant side-effects, not often used
Elemicin	Mace ⎦	
Muscarine	*Amanita muscaria* (fly agaric mushroom)	Mimics some actions of ACH (acetylcholine) throughout the body. Unpleasant side-effects include nausea and vomiting, sweating, cramps and palpitations
Atropine, scopolamine	Deadly nightshade (*Atropa belladonna*), ⎤ Thorn-apple (*Datura stramonium*), mandrake (*Mandragora officinalis*) ⎦	Block the actions of ACH mimicked by muscarine. Have sedative and anaesthetic actions. Side-effects are dry mouth, blurred vision
Scopolamine	Henbane (*Hyoscyamus niger*)	As above
Phencyclidine (Sernyl, PCP, peace powder, angel dust)	Synthetic	A dissociative anaesthetic and analgesic now withdrawn from use as a result of its hallucinogenic side-effects
Ketamine (Ketalar)	Synthetic	Similar to phencyclidine but with lower psychedelic potency

some scientific investigations into the effects of psychedelic drugs in human volunteers. These studies were carried out in the late 1950s on students from medical schools and colleges in North America who were not slow to appreciate the pleasurable properties of the drugs under investigation. The students quickly became keen propagandists for the use of psychedelics. The drug culture which resulted reached a peak in the late 1960s and was centred on the Haight-Ashbury area of San Francisco.

One of the chief apostles of this alternative road to enlightenment was Dr Timothy Leary, who rapidly became a leading figure in the pop music scene. He wrote at the time:

> The LSD experience is . . . about . . . merging, yielding, flowing union, communion. It's all love-making. You make love with candle-light, with sound waves from a record player, with the trees. You're in pulsating harmony with all the energy around you . . . Compared with sex under LSD, the way you've been making love – no matter how ecstatic the pleasure you think you get from it – is like making love to a department store window-dummy.

Although it is clear that Leary was saying, in effect, that LSD could act as an alternative to sex, his writings have tended to encourage people to use LSD as an adjunct to sex, as an aphrodisiac in fact. Some people do use sex as a means of escape from reality, as others use drugs, so it should not be surprising that most of the drugs used recreationally have been reported as acting as aphrodisiacs.

A very revealing survey, published from the Haight-Ashbury Free Medical Clinic in 1982,[17] contained several comments from drug-using patients. Among them one finds, in answer to the question 'Is there a true love drug or aphrodisiac?', the following types of statement:

> 'MDA offers a strong impetus to enhance sensual experience and get close to people you already love.'
> 'Coke makes me higher energy. I'm buzzed, excited – my whole body crackles with sex.'
> 'Grass relaxes me. I'm more sensitive to touch – music and laughter comes through – and love-making achieves a newer and higher plane.'
> 'I can come forever on opium.'

But the most interesting comment received was:

'*My* aphrodisiac might be somebody else's paranoia.'

This reveals an intrinsic appreciation of the social use of drugs for aphrodisiac purposes: that there is no point in achieving a drug-induced high if one's partner is not in the same enlightened state.

Despite these highly subjective remarks by regular drug users, there is very little reliable evidence that any of the psychedelic drugs can consistently, or with certainty, increase sexual pleasure or performance. One of the features of their action is that their effects are very unpredictable and depend entirely on the mental state of the user and the circumstances under which the drugs are taken.

Lysergic Acid Diethylamide

LSD was first synthesized in 1938 by Hofmann, a chemist at the Sandoz Laboratories in Switzerland as part of a project to find potentially useful drugs among the derivatives of ergot, a fungus which is found on rye. Hofmann was also the first person to undergo an LSD 'trip' when he accidentally ingested a minute amount of the drug in April 1943. LSD is astonishingly potent, the minimum effective dose in man being only $10\,\mu g$, a quantity of powder only just visible to the naked eye. It has the advantage of being relatively non-toxic; fatalities among LSD users are most commonly associated with accidental death as a result of the mistaken impression that the drug has given one the ability to fly. It has the disadvantage of producing a rapid tolerance to its effects so that progressively larger doses are needed, but there is no evidence of dependence developing. Grof, reporting the use of LSD in psychotherapy, has noted no effects on sexual behaviour in many couples, but in some there were experiences bordering on the ineffable.

Some drug-induced trips by particular individuals, capable of expressing their dreams or experiences, have yielded some striking pieces of literature. One has only to think of the 'Kubla Khan' of Coleridge, or Huxley's *Doors of Perception*, but the more recent advocates of drug-assisted imagination have produced little of lasting worth. The promiscuity expressed in their work is probably a function of personality rather than of the drugs employed. The uncertainty of the effects achieved by taking LSD renders it a risky undertaking and the drug a highly dubious aphrodisiac.

Other Synthetic Psychedelics

Regular users of LSD often report having tried DMT, MDA, DOM, psilocybin or mescaline as well. As with the reports of the effects of

LSD, there is considerable variation between users. A principal reason for this is that the drugs available on the street are rarely pure and often contain a cocktail of active substances. When both the nature of the drug taken, as well as the dose, are uncertain, it becomes impossible to draw any firm conclusions from subjective user reports. The known abuse potential of all these compounds probably ensures that no controlled clinical trials will ever be carried out, so that it is highly unlikely that any final decision concerning the aphrodisiac potential of these drugs will be feasible.

Hypnotics and Sedatives

These drugs, like the general anaesthetics, are all general depressants which produce a spectrum of behavioural effects from excitement and disinhibition at lower doses to unconsciousness and even full surgical anaesthesia at high doses. The abuse of chlorinated hydrocarbons and other organic solvents (glue-sniffing) is a relatively recent phenomenon, although chloroform and ether were popular in the middle of the last century, and not just for their anaesthetic effects. The object of inhaling these volatile substances is to achieve a sudden 'high'. The narcotic addict achieves it by injecting the drug into a vein so that it is carried rapidly to the brain within seconds. The next most rapid means of transmitting a drug to the brain is to inhale it so that it passes directly across the lung tissues into the pulmonary circulation. The transfer of gases from the lungs to the blood is so efficient that the brain will receive the drug within a few seconds. The smoker who inhales deeply on the first cigarette of the day probably finds that this is the one occasion when the immediate effect of the nicotine can be appreciated.

Volatile liquids such as carbon tetrachloride, trichloroethylene (dry cleaning fluid), chloroform or even alcohol will initially produce a disinhibition of the higher centres in the brain, delirium, and perhaps hallucinations. With general anaesthetics administered under medical supervision, this phase rapidly proceeds into anaesthesia in which the patient is completely inert and unaware of his surroundings. There is also a period of amnesia which will include the time immediately before full anaesthesia was attained. The social user of alcohol, on the other hand, seeks to achieve a prolonged period of disinhibition without the subsequent progression into unconsciousness. But even experienced drinkers have been known to underestimate the anaesthetic potency of alcohol and finish an evening horizontally.

All the chlorinated hydrocarbon solvents are toxic towards the heart, liver and kidneys after prolonged or repeated use. Even chloroform is hardly used today in spite of being one of the earliest and best general anaesthetics discovered.

The early use of chloroform in midwifery was contentious and widely criticized on a variety of religious as well as medical grounds. At a meeting of the Westminster Medical Society in 1849, reported in the *Lancet*, a Mr Gream referred to several cases in which women, under the influence of chloroform, made use of obscene and disgusting language. This fact alone he felt was sufficient to merit the prevention of the use by English women of chloroform. At the same meeting a Dr Tanner reported a case in which a prostitute given ether for an operation claimed to have had lascivious dreams. These apparently immoral and intemperate effects of general anaesthetics were considered by some physicians to be sufficient to justify withdrawing the drugs from use on women. However, a Dr Rogers was able to report that his patients 'prayed most eloquently and sang psalms and hymns in an angelic strain' while being anaesthetized. These cases all tend to illustrate the point that general depressant drugs will act to remove the inhibiting influences of the higher centres and thus unmask the normally latent and perhaps less pleasant aspects of the personality.

The barbiturates, which were at one time used widely as sleeping pills, have been largely superseded by the benzodiazepines, although these too have recently been found to be less than ideal. The popular barbiturates of abuse were taken for their intoxicating effects, and the balance of opinion among users suggests that they tended to decrease sexual performance in the male by making it more difficult to achieve or maintain an erection. This disadvantage outweighed any brief benefit gained from the initial disinhibiting phase of action. Hypnotics and sedatives as a group do not seem to be used for enhancing sexual activity, but there is one remarkable exception, methaqualone or Mandrax. Introduced as a safer alternative to the barbiturates, it rapidly became a cult drug in Europe and America. The subjective verdict of female users was that it had a definite action to increase their libido. Their male friends tended to agree, although in their case it acted like all the other sedative hypnotics and reduced performance. The story of Mandrax has been recounted in detail in Chapter 5.

Even primitive communities have managed to discover naturally occurring substances with soporific or inebriating effects. In South

Africa the hottentots use *kanna*, which is obtained from various species of *Mesembryanthemum*; in the islands of the South Pacific, *kava-kava* is an inebriating drink which plays an important part in social rituals. The drink is prepared from the crushed or masticated roots of a shrub, *Piper methysticum*, which is found on islands throughout the Pacific and the Coral Sea. The preparation and drinking of kava used to be surrounded by many religious and social taboos, but it has since become a preferred alternative to alcohol. It has a two-fold action; it first produces disinhibition so that the consumer becomes more fluent and lively in his behaviour, but without the aggressive tendencies found with alcohol. As time progresses, the legs begin to feel weak, the gait becomes unsteady, and the subjects appear to be drunk. However, kava is also unlike alcohol in so far as there is no unpleasant hangover to follow the period of inebriation. European sailors have reported erotic feelings and increased sexual desire after drinking kava prepared by the South Sea Island women. This is not really so surprising; sailors have an entirely expected tendency towards erotic feelings on meeting members of the opposite sex after long periods of abstinence at sea. The behavioural effects of kava perhaps deserve some further investigation.

Kola, Khat and Other Stimulants

The use of stimulant drugs has become a habit in both civilized and uncivilized races throughout the world. From cocaine and tobacco in the Americas, through tea and coffee in Europe, kola nuts and khat in Africa and the Near East, to betel nut and camphor in the Far East and Pacific Islands, almost every racial group has found some local or imported plant with stimulant properties which it has taken as a keystone into its social culture. Most of these drugs are associated with social rather than sexual intercourse, but several have attracted a local reputation as aphrodisiacs, more probably on the basis of expectation than realization.

Khat is derived from the shrub *Catha edulis*, which grows widely in East Africa and Southern Arabia. The young shoots and leaves are chewed, as with coca, and the behavioural effects are very similar. It produces 'joyous excitation and gaiety', the urge to sleep is banished, and feelings of hunger are diminished. The use of khat pre-dates the introduction of coffee into the Yemen, and until recently there was an open khat market in the port of Aden.

The compulsive use of khat by the local natives in several countries in Africa and Arabia, and its social and economic consequences, provoked the UNO Division of Narcotic Drugs into investigating the dependence potential of the plant.[19] Their studies indicated that the active principle in the young leaf is *dl*-cathinone (α-aminopropiophenone), which is structurally related to amphetamine.

cathinone

amphetamine

The leaves also contain several alkaloids (the cathedulins) and glycosides, but the euphoric and stimulant properties of khat can be explained solely in terms of their content of cathinone. Animal studies have suggested that cathinone is about half as potent as amphetamine and is slightly more specific in its actions. Human studies have been limited to investigations of khat chewers, among whom impairment of the male reproductive system is relatively common. Khat has not yet been taken up to any significant extent by the drug-using communities of America and Europe, but any drug with amphetamine-like effects is bound to have an abuse potential. At present there is no reason to believe that khat or cathinone has any direct aphrodisiac effects.

The kola nut also plays an important part in the social life of the people of North and West Africa. The earliest descriptions of the nut can be found in the writings of El Ghafeky, a Moorish physician of the twelfth century, and the introduction of the kola nut to man is explained in a story of divine intervention rather analogous to the legend of Mama Coca in Peru:

'One day when the Creator was on earth observing the sons of men and busy among them, he put aside a piece of the kola nut which he was chewing and forgot to take it with him when he went away again. A man saw this and seized the dainty morsel. His wife tried to prevent him from tasting the food of God. The man, however, placed it in his mouth and found that it tasted good. While he was chewing, the Creator returned, sought the forgotten piece of kola, and saw how the man tried to swallow it. He quickly grasped at his throat and forced him to return the fruit. Since that time there can seen in the throat of men the 'Adam's apple', the trace of the pressure of the fingers of God.'

Kola nuts are the fruit of a tree, *Cola acuminata*, which grows in West Africa and has been introduced into India, Sri Lanka, Tanzania and the West Indies. The nuts have always been highly prized; at the end of the last century the price of a single nut in the Niger Coast was four shillings, and a slave could be bought for a few nuts. The demand for the nut was not based solely on its pleasurable properties, but also on its widespread use as a status symbol in society.

Chewing pieces of fresh nut produces a sense of alertness and physical well-being, and feelings of fatigue and hunger are allayed. It can also increase the capacity for physical work. The reputation of the nut in West Africa is that it increases sexual performance in men and promotes conception in women. The nut contains about 1.5 per cent caffeine, somewhat less theobromine, and small quantities of kola-red, kolatin and a glycoside, kolanin. The effects of chewing the nuts are almost certainly due to the caffeine, which would be rapidly absorbed through the mucous membranes of the mouth. Extract of kola nut has been, and still is, used in a number of mildly stimulant drinks. Originally the concentration of kola extract was much higher since it provided much of the flavour and colour of the drink, but artificial flavours and colours have since been added and modern cola drinks such as 'Coke' and 'Pepsi' generally contain no more caffeine than can be found in 'Lucozade'.

Caffeine has been the subject of much research over the years and its effects on cell metabolism are well known. Studies on the influence of caffeine on reproduction have been mainly concerned with its mutagenic and possibly teratogenic effects on the fetus. As a thoroughly legitimate drug, and the most widely used stimulant in the world, caffeine would be an ideal drug to test scientifically for aphrodisiac effects. We know that it can increase wakefulness and physical stamina, and that tea, coffee and cocoa (which all contain caffeine) have all been regarded as sexual stimulants in the past, yet there is still no good evidence of a firm foundation for this belief.

8 DRUGS USED CLINICALLY AS APHRODISIACS

Introduction

Any state of ill-health involving mental or physical discomfort is likely to have a detrimental effect on the libido. Impotence and frigidity are often merely symptoms of some underlying disorder which can be readily treated by drugs specific to that disorder. The best example of this is provided by the drugs used in the treatment of depression. The improved state of mind of the patient often coincides with the restoration of the sex drive, but it cannot really be said that the drugs taken by the patient are acting as aphrodisiacs except in a very indirect sense. The current treatment for organic, as opposed to psychogenic, impotence usually consists of hormone replacement; this has already been discussed in Chapter 6. However, until fairly recently, testosterone was sold in combination with yohimbine under the name Potensan for clinical use as an aphrodisiac. Its removal from the British National Formulary marked the end of the era of the formally recognized aphrodisiac drug in clinical medicine.

Despite the loss of aphrodisiacs as a category from the pharmacopoeiae, there are many clinically used drugs which have been reported to have aphrodisiac side-effects. In some cases, such as geriatric patients for example, these side-effects are unwanted and inappropriate, but in others there have been some attempts to make use of the side-effects to improve the sex life of patients. The drugs in question tend to interact with the neurotransmitter systems of the brain, and I have therefore included a brief account of the roles of dopamine and serotonin in sexual function in order to make the mechanism of action of these drugs more clear.

Drugs with positive effects upon the libido are much less common than drugs that reduce the libido and have detrimental effects on sexual function. In some cases these side-effects are so marked that patients become unwilling to continue with the medication once they recognize the cause of their difficulties. Drugs that can affect the ability of the patient to drive a car or operate machinery have an appropriate warning prominently displayed on the packaging, but an equivalent message concerning sexual function is not considered necessary, even though such a warning could prevent much anguish.

The Use and Limitations of Testosterone

It is well established that very large fluctuations in testosterone levels produce changes in the male libido. The effects of castration and the restoration of virility by the administration of testosterone injections are thoroughly documented. However, a great deal of experimental data have been derived from animal studies in which the circulating sex hormones play a much greater role than in man. The influence of these hormones tends to decline as the evolutionary tree is ascended and the neocortical, or thinking areas, of the brain increase in importance. The higher animals show less instinctive sexual behaviour and, in man, the cerebral influences or 'encephalic' elements in the libido probably outweigh the hormonal influences.

Testosterone production by the Leydig cells of the testis is stimulated by secretion of luteinizing hormone (LH) from the pituitary gland. LH secretion is in turn regulated by a releasing factor (LHRH) produced in the hypothalamus. If plasma testosterone levels fall, then LH secretion is stimulated in a form of feedback control loop. Treatment by injection with testosterone as replacement therapy tends to suppress LH release so that endogenous testosterone levels fall even further. The injected hormone is therefore not always as effective as one might expect. Similarly, if the disorder is at the level of the pituitary tissue such that the levels of LH are reduced, treatment with LHRH injections does not always stimulate the release of LH due to more feedback controls that exist. Another type of disorder, at the level of the hypothalamus, in which LHRH is no longer released in sufficient quantity, can result in reduced circulating levels of both testosterone and LH.

Two further endocrinological disorders can lead to functional impotence, namely hyperthyroidism and hyperprolactinaemia (that is, abnormally high blood levels of thyroid hormone or prolactin, respectively). In the latter case, for some reason, testosterone replacement therapy is ineffective and the only successful treatment has involved lowering the prolactin level with drugs such as bromocriptine (see page 225). Prolactin obviously exerts some control over testosterone synthesis or release since spontaneous rises in circulating testosterone levels usually occur at the time when prolactin levels are at their lowest. It should be apparent, therefore, that the hormonal treatment of functional impotence with testosterone alone should only follow a full analysis of the endocrinological state of the patient.

There is still some doubt concerning the extent to which testosterone regulates the level of the sex drive. One theory holds that the sex drive and testosterone level are directly correlated (and in some cases this appears to be true). The alternative theory is that they are inversely related (and there is some evidence for this also). In one fairly well-controlled human volunteer study[28] testosterone levels in six male subjects correlated well with the two measures of their sexual arousal that were made while they watched erotic films. The first measure was the peak erection achieved (estimated by the use of a pressure transducer attached round the base of the penis), and the second was the speed with which a full erection was achieved. On the other hand, there was an inverse relationship between the subjects' testosterone levels and their reported frequency of orgasm prior to the experiment.

Yohimbine

Yohimbine is an alkaloid obtained from the bark of the West African *yohimbéhé* tree (*Pausinystalia yohimbe*, family Rubiaceae). It has featured as an aphrodisiac in African and West Indian medicine for centuries, and it also appears in some modern pharmacopoeiae under the name of aphrodine or corynine. The hydrochloride salt of yohimbine is a constituent in several proprietary preparations with implied aphrodisiac potency.

Pharmacology of Yohimbine

Yohimbine is an α_2-adrenergic receptor blocking agent; that is, it will tend to antagonize the normal activity of the sympathetic nervous system which is mediated by α_2-receptors. There are numerous drugs in this category and most of them are more potent than yohimbine, yet none of them has attracted the same attention or reputation as an aphrodisiac. Drugs of this type have various clinical applications in conditions where an increase in peripheral blood flow is desirable. They are used particularly in the treatment of diseases in which the circulation becomes restricted (e.g. Reynaud's syndrome, phlebitis), for diagnosing catecholamine-secreting tumours, and in the treatment of pulmonary congestion and shock. They have also been employed to prevent cardiac arrhythmias. Yohimbine is more toxic and less specific than many other α-blocking drugs and it is no longer the drug of choice in clinical medicine.

Is it possible that yohimbine has aphrodisiac actions by virtue of its α-adrenergic blocking activity? The development of an erection in the male is initiated by a dilatation of the arterioles in the penis, triggered by neurones which release acetylcholine as transmitter. As the penis swells with the additional inflow of blood, the veins are physically compressed, thus preventing the normal outflow of blood from the organ. In this way the erectile tissue in the penis becomes engorged with blood and an erection is achieved. The loss of an erection is due to the activity of sympathetic nerves which cause constriction of the arterioles. Any α-adrenergic blocking drug which produces vasodilatation by inhibiting the normal vasoconstrictor tone could therefore assist in the initiation and maintenance of an erection. Unfortunately, the actions of drugs of this class are not restricted merely to the blood vessels that supply the penis, and their other peripheral and central effects make their indiscriminate use in this respect a potentially hazardous undertaking.

The effects of yohimbine in man have been reported in the medical literature,[1] and it is probably worth describing the results in some detail, if only to discourage any potential drug user from attempting the experiment upon themselves.

The yohimbine was administered by intramuscular injection to a group of volunteers at a dose of 0.5 mg per kilogram body weight. Their initial symptoms consisted of perspiration, salivation, lachrymation and pupillary dilation while, at the same time, the blood pressure and heart rate were increased. Some subjects reported feelings of nausea and an urge to urinate or defaecate. Between 10 and 20 per cent of the male subjects experienced an erection, although they were not in any state to really enjoy it. It is probably this last effect which is responsible for the reputation of yohimbine as an aphrodisiac.

In addition to the peripheral symptoms listed above, the subjects also reported various psychological effects, notably feelings of tension, anxiety and irritability. Some experienced restlessness and nausea, and the group generally expressed their unwillingness to repeat the experiment. The dose they received was not likely to be sufficient to produce any significant blockade of the α-receptors, and the symptoms they exhibited suggest a direct cholinergic stimulating action in the periphery. It is interesting to note that the induction of fear and anxiety by injections of yohimbine and similar drugs has been used both as a form of torture and as a means of testing new drugs for any potential anti-anxiety effects.

Yohimbine and Testosterone Preparations

Afrodex is the trade name for a preparation marketed in the US consisting of extract of nux vomica (strychnine) 5 mg, methyl testosterone 5 mg, and yohimbine hydrochloride 5 mg. It is widely considered to be of value in the treatment of male impotence. A series of reports in the medical literature have described over 4000 cases in which Afrodex was prescribed to male patients complaining of some degree of impotence and ranging in age from 20 to 91 years.[2,3] The results were obtained by asking the patients to report weekly, for a period of ten weeks, whether or not they felt any change after taking the preparation. They had to describe their responses as excellent, good, fair or poor. Over the ten-week period the proportion of patients reporting a poor response declined from 40 per cent to 4 per cent, and those reporting excellent increased from 10 per cent to 50 per cent.

This can hardly be considered a well-controlled or even scientific study. There were no placebo controls (i.e. patients given identical but inert capsules), and the method of self-assessment was totally subjective and extremely vague. Looking more closely at the data from this study it becomes clear that progressively fewer patients were reporting each week. Assuming that a patient would be more likely to report if he felt some positive benefit, then it is clear that the responses would be increasingly unrepresentative of the group as a whole.

Two points did emerge from this study, however, firstly, that the effect, if any, of the Afrodex was not apparent until at least two to three weeks after the start of the medication; and secondly, that the treatment was markedly less effective in patients over 60 years of age.

More recent and better designed studies have since been reported.[4,5] In a double-blind crossover trial, Afrodex was tested against an inert placebo made up in an identical capsule. Half the patients received each treatment for four weeks. A four-week rest period following during which no medication was given, and then the two groups received the alternative therapy for a further four weeks. Neither the patients nor the physicians actually distributing the capsules were aware which group of patients was receiving the Afrodex until the trial had been completed. This double-blind crossover design is still one of the best methods for testing the effectiveness of a particular drug treatment.

Rather than record subjective self-assessments, the patients were asked to record the number of erections and orgasms that they achieved over each fortnight throughout the trial period. The patients were also divided into categories based upon the origin and duration of their impotence. These categories included psychogenic impotence, alcoholic impotence, hypogonadism, and so on. The results are summarized in Table 8.1.

Table 8.1: Effects of Placebo versus Afrodex on Sexual Activity

	Start	4 Weeks' Placebo	Change (%)	Start	4 Weeks' Afrodex	Change (%)
Total number of erections per week (21 subjects)	3.25	11.25	+ 246	4.25	49.5	+ 1065
Total number of orgasms per week (21 subjects)	3.25	8.75	+ 169	3.75	23	+ 513

The results are quite spectacular and indicate the importance of carrying out proper controls, since the supposedly inert placebo has produced a very marked improvement in performance. This effect has to be subtracted from the effect of the Afrodex in order to determine the response due to the drug itself. In the results obtained from the group of patients who had the Afrodex first, it appears that the effects of the drug were maintained for up to four weeks after ceasing treatment (Table 8.2).

Table 8.2: Effects of Afrodex versus Placebo on Sexual Activity

	Start	4 Weeks' Afrodex	Change (%)	Start	4 Weeks' placebo	Change (%)
Total number of erections per week (27 subjects)	6	67.25	+ 1025	15	11.75	−22
Total number of orgasms per week (27 subjects)	4.25	39.25	+ 823	11.5	9.25	−20

Under these conditions the placebo effects have been eliminated, and the effects of the Afrodex are more clear cut. The results in this study also showed that the Afrodex required between one and two weeks to produce its maximum effect, although the improvement was then maintained beyond the end of the trial. It would be interesting to know which of the three constituents of Afrodex was most responsible for this remarkable therapeutic effect on sexual performance. The pharmacological properties of yohimbine have already been described, and strychnine also seems to be an unlikely candidate (see below). On balance, and particularly in view of the long latency to onset of the effects, it would seem that testosterone is most likely to be responsible for the effects observed. This view is also supported by the fact that many of the patients would have had abnormally low testosterone levels initially, and there is no evidence that Afrodex (or any similar preparation) is capable of increasing sexual potency above the normal non-pathological level.

Two preparations which are approximately equivalent to Afrodex have been available in the UK. They are: Potensan, which consists of yohimbine hydrochloride (5 mg), strychnine hydrochloride (0.5 mg), and amylobarbitone (15 mg) in tablet form, and Potensan Forte which contains yohimbine hydrochloride (5 mg), methyl testosterone (5 mg), pemoline (10 mg) and strychnine hydrochloride (0.5 mg). The latter was advertised in the medical literature as the specific treatment for lack of libido in either sex, the indications being impotence, frigidity, sexual fatigue, and so on. However, no specific claims were made as to its efficacy and it has now disappeared from the British National Formulary.

Potensan was subjected to a controlled clinical trial for the treatment of impotence by a group of workers in Edinburgh.[6] In their study the urinary testosterone levels were measured in all the subjects prior to the start of the trial and found to be about 37 per cent below normal. The design of the trial was similar to that outlined above, but the results were far less positive. However, the mean incidence of successful coitus leading to orgasm and ejaculation was significantly increased from 0.12 times per week during the control period to 1.12 times per week after four weeks of treatment with Potensan Forte (one capsule four times a day).

There was no indication that the patients with the greatest initial testosterone deficiency experienced the greatest improvement, which suggests that the effects of Potensan may not be due merely to the replacement of insufficient testosterone. Also, in contrast to the studies with Afrodex, no long-term improvement was observed, and several of the subjects suffered a deterioration in their sexual performance one year after the Potensan Forte trial, although it is not clear from the report whether, at that later stage, they were still taking the capsules.

The dose of yohimbine and methyl testosterone is the same in both Potensan Forte and Afrodex, and apart from the presence of pemoline (a very mild stimulant of doubtful efficacy), the only difference between the two preparations is the concentration of strychnine, which is ten times higher in Afrodex. If we assume that the groups of patients used in the trials of the two preparations had similar clinical symptoms, and if the difference in the apparent potency of the two preparations was not due to the methods of assessment employed by the different groups, then it would appear that it is the strychnine which is contributing to the significant effects of Afrodex. Certainly strychnine does act as a stimulant or 'tonic' at low doses, but the safety margin between the therapeutic dose and the lethal convulsive dose is very narrow (see below).

It is perhaps worth noting that the Edinburgh group also recorded a marked improvement in their placebo-treated patients, the frequency of orgasm being 0.56 times per week in those groups. These two independent studies are therefore in good agreement on one important observation, namely that a placebo can be an effective treatment for a diminished libido. The reputation of the majority of the aphrodisiacs described in this book probably rests on this fundamental observation.

Strychnine

Strychnine is the principal alkaloid present in the large seeds of the nux vomica tree (*Strychnos nux-vomica*), which also contain brucine, an alkaloid with similar properties to strychnine but less potent. It is now recognized that strychnine has no therapeutic value, although at one time it was included in a large number of 'tonic' preparations. The main action of strychnine in the body is to block the inhibitory neurones which release the amino acid glycine as a transmitter. These neurones normally exert a feedback control system on the motor neurones which branch from the spinal cord and control voluntary muscular movement. Consequently, strychnine acts as a potent convulsant, producing co-ordinated muscular spasms similar to tetanus. The subject becomes hypersensitive to any sensory stimuli such as light, sound or touch which can then trigger a muscle spasm. Death occurs as a result of spasm of the muscles controlling respiratory movements.

Strychnine poisoning still occurs occasionally, particularly in children, since it is used as a rodent poison in the form of small impregnated biscuits which can be attractive to children. Strychnine is rarely used as a means of committing suicide because of the unpleasant and painful symptoms that precede respiratory paralysis and death. Nevertheless, strychnine was long regarded as a valuable tonic, the reasons for this being that it has an extremely bitter taste (supposed to stimulate the appetite), and that it increases muscle tone. However, the safety margin between the dose required to increase muscle tone and the toxic dose is very small indeed, and there are now much safer alternative tonic bitters available.

Strychnine as an Aphrodisiac

Dr Edward Martin, writing in *A System of Therapeutics* in 1892, discusses the treatment of male impotence and sterility and recommends the use of a general tonic containing nux vomica, dilute phosphoric acid and Huxham's tincture. The combination of phosphorus, damiana and strychnine is also recommended for sterility, although the author admits to never having used the drugs himself. He also mentions that cantharides and alcohol in low doses have distinct aphrodisiac effects. For the specific treatment of aspermatism (defined as lack of an emission during intercourse) due to lack of tone in the muscles responsible for ejaculation, he provides the following prescription:

℞	Strychnine sulphatis	gr 1/40	(1.5 mg)
	Phosphori	gr 1/200	(0.3 mg)
	Extract damianae	gr iij-M	(0.25 ml then mixed)

The mixture is to be made into pills, two to be taken three times a day. Dr Martin points out that this will not be effective in cases where the inability to ejaculate results from a diminished sensitivity of the penis to tactile stimulation. Under these circumstances he suggests an 'electric brush' but unfortunately gives no details of this intriguing device.

Luteinizing Hormone Releasing Hormone (LHRH)

LHRH was isolated from pig hypothalamus in 1971 and found to be a potent releaser of both LH and follicle stimulating hormone (FSH) from the pituitary. It was thought initially that LHRH might prove to be of value in treating the infertility associated with reduced LH and FSH levels. For various reasons this hope has not been fulfilled, although paradoxically it now seems that LHRH or a synthetic analogue could be useful as a contraceptive in women. Animal experiments indicated that LHRH induced mating behaviour in female rats, but there is no evidence of an equivalent effect in the female of our species.

One of the earlier investigations into the use of LHRH in male patients with hypogonadism resulting from hypothalamic or pituitary disease was published in 1974.[7] Twelve patients were given a short

course of treatment consisting of 500 μg of LHRH injected subcutaneously every 8 hours. As expected, the circulating levels of LH
and FSH in the blood rose rapidly and, in the adult patients, sexual
potency was restored and sperm production (assessed by regular sperm
counts) recommenced within 7 to 14 days of starting the treatment.
The best dose regimen appeared to be chronic intermittent courses of
injections since, with prolonged use, LHRH tends to become less effective as the normal feedback controls are activated. The disadvantages of
this form of treatment for impotence are that it requires regular self-
administered injections, with the attendant risks of infection, and that
LHRH is also extremely expensive. In the UK the current cost of the
course of treatment described above would be over £40 a day.

LHRH has also been tested in the form of a nasal spray in a double-
blind trial with 20 male patients suffering from sexual impotence.[8]
Their average age was 47.9 years and the treatment consisted of 1 mg
per day for a period of four weeks. A greater therapeutic effect was
found in the LHRH group as compared with the placebo group. The
beneficial effects persisted after the end of the trial for 4 to 6 weeks.
This limited trial suggests that the problem of self-administered injections could be overcome, but the cost would, at present, be even more
prohibitive.

LHRH has been produced commercially by synthetic procedures and
has the generic name Gonadorelin. It is used as a diagnostic test for
assessing pituitary function, but does not seem to have proved sufficiently effective or free from side-effects in the long-term treatment of
male impotence. It is possible that a new synthetic analogue of LHRH
will be developed which could be valuable in restoring the libido of
hypogonadal patients on a permanent basis at reasonable cost.

Dopamine and Sexual Function

Quite often in clinical pharmacology, a drug which has been prescribed
for a particular disease is found to produce side-effects which turn out
to be of benefit to a completely different group of patients. Many
modern drugs now in use are a direct consequence of such fortunate
discoveries. The side-effects of *L*-DOPA in Parkinsonian patients
provided a clue to the discovery of the involvement of the natural
transmitter dopamine in the regulation of sexual function and libido.
Specific nerves containing dopamine in their terminals run from the
hypothalamus to the anterior pituitary in the brain. They release

Figure 8.1: Sequence of Events which Regulate the Release of Prolactin

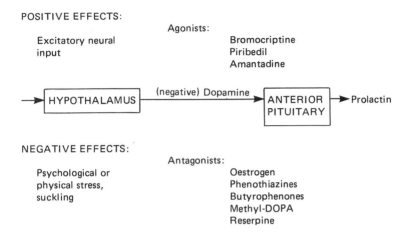

dopamine which acts to suppress the secretion of the hormone pro-lactin. Many cases of male impotence, female menstrual disorders, and infertility in both sexes have been shown to be associated with abnormally high circulating levels of prolactin in the blood. The role of prolactin in regulating sexual function is described in Chapter 6. The rational treatment in cases of this sort would seem to be the reduction of prolactin levels to normal. This cannot be achieved by the administration of dopamine itself since it is rapidly destroyed by enzymes before it can reach its site of action in the brain. However, it was postulated that drugs which mimicked the action of dopamine, and were already in use in the treatment of Parkinsonism, might well be of use in relieving the symptoms of impotence and infertility result-ing from excess prolactin. The sequence of events which regulate the release of prolactin are shown in Figure 8.1. The positive effect of dopamine-like drugs (agonists) can be seen to counteract the negative effects of stress or dopamine antagonists.

Bromocriptine

Bromocriptine acts like dopamine to inhibit the release of prolactin. It was introduced in about 1971 for the treatment of hormonal disorders involving both prolactin and growth hormone where alternative treat-ments had failed. The limited use of bromocriptine has been due firstly to its cost (about £5 to £10 per day), and secondly, to a high incidence of side-effects at the larger doses needed for treating Parkinsonism. Nevertheless, controlled studies with bromocriptine have indicated that it can restore normal sexual function in cases where the underlying cause is oversecretion of prolactin.

One particular example involved 47 male patients aged between 21 and 58 years, of whom 43 were suffering from sexual impotence and 4 of whom experienced premature ejaculation.[9] Bromocriptine proved beneficial in roughly half the cases and many of these also showed a reduction in circulating prolactin levels. The subjects were interviewed and asked specifically about their libido, spontaneous erection frequency, ejaculations, masturbation, intercourse and orgasm, both before and after treatment. The results, compared with those for a placebo preparation, are shown in Table 8.3.

Table 8.3: Effect of Bromocriptine versus Placebo on Sexual Function

Treatment	Good Effects (normal sexual func- tion restored)	Slight Effect	No Effect
Bromocriptine (2.5 mg three times a day)	52%	22%	26%
Placebo (inactive)	44%	31%	25%

Although the results with bromocriptine look extremely promising, it is very significant that the supposedly inactive placebo treatment with which it was compared also had a marked beneficial effect. Other studies have proved more conclusive but this one is useful in that it demonstrates the problem of eliminating the positive influence of medical attention in itself.

At the dose of bromocriptine normally prescribed for this type of hormonal disorder, there have been very few serious side-effects reported, mainly feelings of nausea, dizziness, lethargy, heartburn and constipation. It does not appear to have the same hallucinatory or delusional effects of *L*-DOPA, and in Parkinsonian patients there have been no reports of abnormally increased libido or erotic dreaming in octogenarians.

In spite of the responsible medical reporting of the results of controlled studies with bromocriptine in male patients with prolactin-induced impotence, the popular press soon headlined it as the 'sensational sex drug'. The misleading nature of such reporting naturally caused many people to believe that bromocriptine could restore sexual function in all cases of impotence as well as increase the sexual capacity of normal individuals. Once again, the myth of the aphrodisiac, like the philosophers stone of the middle ages, caused normally responsible, rational people to suspend their critical faculties.

Other dopamine agonists such as amantadine and piribedil have not attracted quite the same attention, presumably because they no longer represent a novel class of drug. To conclude, it is worth emphasizing that in normal males, with a normal hormonal balance, bromocriptine has no effect on the libido and only the side-effects will be apparent. Although a raised prolactin level can produce a functional impotence and diminished libido, there is no evidence that an artificially lowered prolactin level has the opposite effect.

L-DOPA

L-DOPA or *L*-dihydrophenylalanine is one of the most effective treatments available for relieving the unpleasant symptoms of Parkinson's disease and has been in use for about 15 years. Parkinsonism is a progressive degenerative disorder which tends to occur in patients over 40 years old. It causes a loss of the fine control of movement, particularly walking, and a tremor of the hands. It involves a specific nerve centre in the brain (the substantia nigra) in which the chemical dopamine acts as a neurotransmitter or local hormone, communicating signals between adjacent nerve cells. Patients with Parkinsonism have a much reduced concentration of dopamine in the nerve cells in this particular area of the brain. Dopamine itself cannot be used for treatment of the disease in the way that a vitamin deficiency can be made up by adding extra doses of the vitamin to the normal diet, because it is very rapidly destroyed by enzymes present in the gut and liver. Consequently, a dose of dopamine taken orally would never reach the brain where it could help to restore the normal level of the transmitter. However, since the brain normally manufactures dopamine from *L*-DOPA, this latter drug was considered to be the rational treatment for Parkinsonism. The advantage of *L*-DOPA therapy is that it can be combined with another drug which prevents it from being destroyed in the body before it reaches the brain.

Over a number of years it has become clear that, although *L*-DOPA can successfully alleviate the symptoms of Parkinsonism, it does have a rather high incidence of side-effects. These side-effects depend very much upon the dose of the drug given as well as on the individual patient. Among these side-effects there has occurred the occasional instance of an obviously increased libido.

The first review article in the medical literature which drew attention to this problem was published in 1969 just a few years after the drug came into regular use in clinical practice.[29] The author reviewed over 100 reports in the literature in addition to information derived from 86 of his own patients (62 men and 24 women). The overall incidence of side-effects was as follows:

	%
abnormal muscular movements	50
nausea	44
low blood pressure (fainting)	31
confusion, hallucinations and vivid dreams	16
depression	11
palpitations	7.5
anorexia, loss of weight	5
somnolence, sedation	5

In this list of side-effects there was no mention of increased libido, but the author did state at a later point in the article:

Personality changes occurred in a number of Parkinsonians. The general mood-elevating and awakening effect of *L*-DOPA was observed in most patients. However, this was accompanied by a false sense of omnipotence and 'insouciance'. The judgement of some patients in the face of major decisions was often faulty. A behavioural pattern with frontal lobe overtones was evident in some patients, especially in the sexual sphere. A clear-cut, visually evident increase in libido occurred in at least 4 male patients but, unfortunately, erections were not sustained and copulation was terminated with premature ejaculation. It is difficult to evaluate the presence or absence of this effect in our female patients, but we believe that it is present in them as well.

A very large number of people have obviously shared this belief subsequently, since this modestly worded statement led to an enormous surge of public interest in the drug, particularly in the United States where promoters and advertisers were quick to seize upon *L*-DOPA as a source of titillation to the gullible. Figure 8.2 from the *Los Angeles Times* is a good example of the genre.

Subsequent clinical studies have examined the potential aphrodisiac

Figure 8.2: An Advertisement for an 'Adult' Film from the Entertainment Section of the *Los-Angeles Times* of 1970. This sensational revelation of the stimulant properties of the anti-Parkinsonism drug *L*-DOPA was based on the flimsiest of scientific evidence

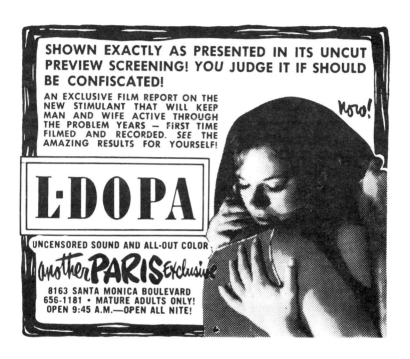

effect of *L*-DOPA in more detail. A report in the *Lancet* in 1970[10] indicated that 18 out of 90 patients with Parkinson's disease experienced mental disturbances while on *L*-DOPA therapy. A breakdown of these disturbances showed that 7 were moderately depressed, 4 were suicidal, 5 were overtly psychotic, 1 was psychotic and depressed, and just 2 out of 49 male patients reported increased sexual drive. As in all studies on Parkinsonian patients, the average age was relatively high, and in this case was over 65 years. Under these circumstances the women patients would all be post-menopausal and the males would be expected to be experiencing a reduced sexual drive purely as a result of their advancing years.

Later studies have specifically concentrated on the sexual aspects of the effects of *L*-DOPA. Bowers *et al.*[11] in 1971 studied 19 patients who consisted of 12 males aged between 45 and 80 years, and 7 females aged between 49 and 74 years. All of them had been taking *L*-DOPA for their Parkinsonism over periods of between 3 and 15 months previously. Their sexual behaviour was rated as follows:

0 :   sexual feelings not present
1 :   some occasional awareness of sexual feelings
2 :   mild to moderate sexual feelings
3 :   occasional intercourse or moderate sexual feelings in the absence of the partner
4 :   intercourse at least every 2 weeks or strong sexual feelings

The average score for this group of patients before the start of the *L*-DOPA treatment was 1.7, and afterwards 2.0. Of the 19 patients, 7 reported activation of their sexual behaviour (6 males and 1 female). No one improved by more than one point on the scale. Also, the effects were not long-lasting and tended to disappear as the therapy continued.

Prior to treatment the symptoms of Parkinsonism might be expected to have a detrimental effect on the sex life of the patients, and some improvement in sexual behaviour could be expected since most patients experienced a remission of their symptoms with the *L*-DOPA treatment. Their feelings of general well-being also increased. A specific stimulation of the sexual drive was noted in only three of the male patients; in one case an 80-year-old man began to experience nocturnal emissions and erotic dreams.

A more recent study[12] in 1978 included some endocrinological information as well as psychiatric interviews and the rating of sexual behaviour. In a group of 7 male patients (a rather small group for this type of study) 5 out of the 7 reported reduced sexual desire following the onset of the disease, and the remaining 2 reported satisfactory levels of desire and potency. Of the 7, two showed increased sexual functioning during *L*-DOPA therapy and also increased blood levels of luteinizing hormone (LH). The significance of this particular hormone in determining the level of the libido has already been discussed (see page 223). A further 2 patients reported increased sexual functioning after previously having a history of impotence. On the other hand, 2 out of 7 reported reduced sexual functioning with *L*-DOPA and one patient felt that he experienced a marked improvement when taking the supposedly inactive placebo preparation.

It is interesting at this point to consider the likelihood of an aphrodisiac effect occurring during *L*-DOPA therapy in Parkinsonian patients.

The overall incidence of increased sexual functioning seems to be about 4 per cent in studies where the patients were not specifically questioned about their sex life, and about 30 per cent when they were. The incidence in male patients is about three times that in females.

All the investigations mentioned so far have been carried out on sick patients from a fairly narrow age group, who were, in many cases, being treated with other drugs at the same time. Bearing this in mind, it is worth asking what the effects of *L*-DOPA might be in (a) healthy subjects who are not suffering from Parkinsonism, and (b) younger people who might be expected to have a stronger sex drive in the first place.

A report[13] of the use of *L*-DOPA in depressed patients (you will recall that one 'side-effect' of *L*-DOPA was to improve mood and engender feelings of well-being) made no mention of any increase in sexual function. However, the doses of *L*-DOPA were rather less than those given to many of the Parkinsonian patients. The incidence and severity of the side-effects of *L*-DOPA tend to increase with increasing dosage. A single case report[14] of a 12-year-old boy given *L*-DOPA in an attempt to improve his epilepsy and other behavioural disorders described strong hypersexual behaviour. After 10 months of therapy at a relatively low dose, there was a very obvious enlargement of his external genitalia. The drug was withdrawn and his genitals stopped growing, but unfortunately the original behavioural disorders returned. This study would have been of far more interest if it had been possible to measure the blood levels of growth hormone, testosterone and luteinizing hormone. *L*-DOPA is known to increase the secretion of growth hormone, particularly in children and adolescents.

L-DOPA has never been proposed as a treatment for sexual dysfunction or loss of libido on an official basis, and the very high incidence of side-effects would make it too risky for use outside Parkinsonism where it is of definite benefit. The incidence of reported aphrodisiac side-effects suggests that it is the background condition of the patient that is important in determining whether or not an increased libido is manifested. Also, it may be considered inappropriate that octogenarians should exhibit renewed sexual awareness and interest. As a direct result of the extravagant reports of the effects of *L*-DOPA that have appeared in the popular press, there has been the occasional instance of the misuse of *L*-DOPA by drug-orientated groups, particularly in the United States. That this has not become a significant problem suggests that the action of the drug is not particularly satisfactory or pleasurable.

Anaphrodisiac Dopamine Antagonists

Since the drugs that act like dopamine can restore sexual function, it is natural to ask whether the drugs that can antagonize dopamine will produce an increased prolactin level and impotence or loss of libido. The phenothiazines and butyrophenones are groups of drugs used widely in the treatment of psychoses and related mental disorders. As major tranquillizers they tend to quieten disturbed patients and reduce their interest in sexual activity. The patients given these drugs are usually exhibiting highly disturbed or abnormal behaviour, and the possible loss of libido is hardly a problem of concern to clinicians when judged against the reduction of psychotic symptoms. The drug benperidol is specifically indicated for the treatment of cases of deviant and antisocial sexual behaviour.

Loss of libido is more a problem when it occurs as a side-effect of drug therapy not intended to affect the mental state or behaviour of the patient. A particular example occurs in the treatment of high blood pressure (hypertension) with methylDOPA or reserpine.[15] MethylDOPA is a close chemical analogue of *L*-DOPA and is metabolized in the brain to methyl dopamine. This metabolite competes with the natural dopamine and, being less active, causes a reduction in dopamine function by competing with the normal transmitter. Although this interaction is not the primary aim of methylDOPA therapy, it results in an anaphrodisiac effect in a proportion of patients taking the drug. Of 22 male patients with high blood pressure, 7 reported loss of libido, an inability to maintain an erection and difficulty in ejaculation during chronic treatment with methylDOPA.[16] Their sexual functions returned to normal after the drug had been withdrawn.

Reserpine is a drug with a long and distinguished history. It is extracted from the roots of *Rauwolfia serpentina* (snake root) and has been used for centuries in Hindu medicine as a cure for mania in general and snakebite in particular. The initial rationale for the use of the drug for snakebite probably lay in the resemblance of the plant root to a snake. It has a calming or tranquillizing effect on disturbed patients and was introduced in about 1950 for the treatment of schizophrenia. It is currently used only for reducing high blood pressure and is the only representative in modern western medicine of a large Indian tradition of herbal remedies. Reserpine acts to deplete the stores of dopamine from the nerves in the brain, and the most troublesome side-effect is depression and apprehension in patients. It also has several endocrinological actions including inhibition of menstruation, reduced fertility and impotence. Although these actions only affect a small proportion of patients, they are probably a direct result of the depletion of dopamine.

There are numerous species of *Rauwolfia* and it is interesting that a decoction of the African serpent wood (*Rauwolfia vomitoria*), when given as a warm infusion by means of an enema, is claimed to be an aphrodisiac by the Gouros of West Africa. This seems rather unlikely, except perhaps in individuals who derive erotic pleasure simply from the administration of an enema. It is more likely that the plant has been confused with a species of *Aristolochia*, which is also known as snake root. These plants are found in most tropical countries and were originally used to aid childbirth. (The Greek name is derived from *aristos*, 'best' and *locheia*, 'childbirth'.) *Aristolochia cymbifera*, which grows in Mexico and Brazil, has a long-standing reputation as an aphrodisiac in local medicine.

Serotonin and Sexual Function

Serotonin (5-hydroxytryptamine, 5-HT) is, like dopamine, a local hormone secreted by specific nerve cells or neurones in the brain in order to communicate signals between adjacent neurones. 5-HT neurones appear to be involved in the control of sleep patterns and the initiation of sleep itself, as well as having some influence on mood. Various theories have been put forward to explain several mental disorders including depression, insomnia, migraine and schizophrenia in terms of disorders of the 5-HT-containing neurones in the brain. The evidence for these theories has been obtained from patients suffering from one or more of these disorders, and in whom altered levels of 5-HT or its metabolites have been detected by chemical analysis.

Animal experiments have also been carried out in which the 5-HT system has been artificially stimulated or inhibited by chemical means and in which the ensuing changes in behaviour have been observed and analysed. It was from experiments of this type that the first report appeared of a drug which depleted the 5-HT in the brain and, at the same time, produced excessive, repetitive sexual behaviour in male animals. This hypersexuality was frequently of a homosexual nature, a very rare phenomenon in animals. It did not take long for the potential human application of such drugs as aphrodisiacs to be appreciated. Although several clinically useful drugs related to 5-HT have been developed, their effects on human sexual behaviour have been rather disappointing after the astonishing effects observed in the early experiments in rats, rabbits and cats.

In order to understand how the 5-HT system can be manipulated, it will be helpful to see the complete metabolic pathway for the synthesis and breakdown of 5-HT:

$$\underset{\text{(1)}}{L\text{-tryptophan}} \longrightarrow \underset{\text{(2)}}{5\text{-hydroxytryptophan}} \longrightarrow \underset{\text{(3)}}{5\text{-HT}} \longrightarrow 5\text{-HIAA}$$

L-tryptophan is an essential amino acid normally present in the diet. The three chemical reactions shown are all catalysed by enzymes found in the body. The first reaction can be inhibited by the drug parachlorophenylalanine (PCPA) and it consequently produces a fall in the 5-HT level. This is the drug that produced hypersexuality in the animal experiments. The effects of PCPA can be overcome by adding large doses of 5-hydroxytryptophan to the diet. This is the immediate precursor of 5-HT and will restore the normal level of 5-HT. 5-HT itself cannot be given since it is rapidly destroyed by enzymes in the intestinal wall so that very little enters the blood circulation and reaches the brain. It also has the property of making the intestinal muscles contract so that in large doses it could produce some quite unpleasant symptoms. One further way in which 5-HT function can be augmented is in blocking the enzyme responsible for its breakdown to 5-HIAA, which is biologically inactive. This enzyme, monoamine oxidase (MAO) also inactivates dopamine and noradrenaline so that inhibiting it with drugs will affect all three transmitter hormones. Nevertheless, MAO inhibitors are used widely to treat depression. These drugs have occasionally been reported to increase libido but this may be an indirect action occurring as a result of the patient recovering from the depression.

There is one more group of drugs which antagonize the actions of 5-HT directly. The first of these to be discovered was the famous hallucinogen LSD (lysergic acid diethylamide). This drug was found to be too dangerous for clinical use, but other safer 5-HT antagonists have since been developed.

5-HT is believed to influence sexual behaviour by controlling the release of gonadotrophin hormones from the pituitary:

$$\underset{\substack{\text{5-HT, Neural} \\ \text{input (+ ve)}}}{\xrightarrow{\hspace{2cm}}} \text{HYPOTHALAMUS} \xrightarrow{\text{LHRH (+ ve)}} \text{PITUITARY} \begin{array}{l} \rightarrow \text{LH} \\ \rightarrow \text{FSH} \end{array}$$

There have been many conflicting reports on the importance of 5-HT in regulating the secretion of LHRH, which, in turn, provokes the release of LH and FSH. There are obvious differences between species and it seems that 5-HT exerts a far more important control over sexual function in lower animals than it does in humans. Nevertheless, drugs that interfere with the 5-HT system have several clinical applications and there have been occasional reports of aphrodisiac side-effects. Since high circulating levels of LH are associated in man with a reduced libido, the scientific prediction would be that 5-HT itself or drugs that augment its action should be anaphrodisiac whereas 5-HT antagonists or depleters should be aphrodisiac. However, nothing is ever quite so clear cut as that, as the following reports will show.

L-*Tryptophan*

L-Tryptophan is used in the treatment of depressive illness on the principle that it will increase the level of 5-HT in the brain. For the same reason it has been tried for the relief of migraine. The earliest report of the effects of *L*-tryptophan in normal subjects was in 1962.[17] Volunteers of both sexes were given doses of between 30 and 90 mg per kilogram of their body weight. The lowest dose produced feelings of drowsiness and fatigue and at the higher doses the subjects tended to fall asleep spontaneously. On being awakened they reported feelings of euphoria, became hyperactive, and generally behaved in an uninhibited way. No obvious sexual interest or awareness was reported.

A later study in 1966[18] involved healthy male volunteers who took 5 to 10 g of *L*-tryptophan immediately before retiring to bed in a sleep research laboratory (these doses are probably slightly higher than those used earlier). The tryptophan produced no significant changes in the sleep patterns of the subjects, although the onset of sleep occurred more rapidly. On waking, the subjects reported feelings of slight euphoria, drunkenness and drowsiness. The investigators noticed that at this time the subjects' conversation tended to be of a lewd and libidinous nature. Whether the lewd thoughts were directly due to the tryptophan or were merely a manifestation of normal healthy male behaviour is impossible to determine. However, a subsequent report gave an account of a group of female patients diagnosed as schizophrenic who were given *L*-tryptophan together with an MAO inhibitor antidepressant.[19] The aim of the treatment was to improve the mental state of the patients who were lethargic, pleasureless and depressed. Five out of the seven so treated showed marked signs of increased sexual interest and awareness. This progressed to the stage where they

began sexually assaulting the male cleaners on the ward. This compulsive sexual excitation also manifested itself in the form of masturbation and homosexual activity. Before jumping to conclusions, it should
be noted that all these patients were taking, or rather, were being given,
other classes of drugs (mainly antipsychotics and tranquillizers) for the
treatment of their schizophrenia symptoms. Also, their libidos were
abnormally low before the start of the experiment and they had all
been hospitalized for some years. The striking effects observed here are
unlikely to be due to tryptophan alone, and certainly would not be
expected or be capable of being repeated in the population as a whole.

The prolonged correspondence that appeared in the *British Medical
Journal* in the mid 1970s mentioned several individual case reports of
depressed or psychotic patients given *L*-tryptophan (usually in con-
junction with other drugs) who then exhibited increased libido and
sexual activity. These effects seemed to be dependent on the dose given
and, in most cases, a reduction in dosage restored their libido to normal
without loss of the therapeutic action of the drug.

For the sake of demonstrating the conflicting evidence available on
the subject of 5-HT it is worth mentioning two interesting case reports
from Italy.[20] In the first, a 42-year-old university professor was treated
with *L*-tryptophan for his intense and persistent headaches. During the
treatment he experienced a lack of orgasm and inability to ejaculate.
He apparently gave up taking the tryptophan for short periods in order
to restore his normal sexual activity. The second subject was a 48-year-
old clinician who also took tryptophan for the relief of severe headaches. To quote the original report:

> This colleague reported a vigorous sexual activity since adolescence:
> in the last two years sexual compulsion increased, sometimes it
> became so intense as to be intolerable; in fact it had become, when it
> was possible to have more than one intercourse in 24 hours.

The tryptophan therapy prevented the occurrence of the headaches
and reduced the patient's urge for intercourse to about two or three
times a week. Although people are notoriously diffident about their
sexual activities, indulging once a day when circumstances permit does
not seem excessive or compulsive. With the cessation of treatment, both
the headaches and the 'hypersexuality' returned. The dose of tryptophan used in these cases was 4 g per day, somewhat less than the dose
used in the sleep experiments.

Theory would predict that *L*-tryptophan should be anaphrodisiac, and at least its use at low doses to treat headaches in otherwise mentally normal subjects tends to support this view. Its aphrodisiac effects only seem to appear in psychotic subjects taking other medication simultaneously, and are most likely to be due to drug interactions.

Parachlorophenylalanine (PCPA)

This drug, which acts to deplete 5-HT, would be expected to have the opposite effects from *L*-tryptophan. The first report[21] of its effects in healthy subjects appeared in 1966 when 6 normal prison inmate volunteers were given gradually increasing doses of PCPA from 1 mg per day up to 3000 mg per day over a period of several weeks. At the highest doses their blood 5-HT levels were reduced to between 60 and 70 per cent of control, and symptoms of tiredness, dizziness, nausea, uneasiness, headache and constipation were reported. There was no mention of any increase in libido or sexual awareness, but it is possible that the volunteers were not asked about this and, in any case, prison inmates are not likely to be in a position to indulge in heterosexual activity and might be wary of admitting to any other means of obtaining sexual gratification.

Subsequent studies confirmed the relative safety of PCPA in humans, but still gave no indication of any aphrodisiac effects. PCPA is used in the treatment of carcinoid syndrome in which an intestinal tumour can be actively synthesizing and secreting 5-HT. The patients have high circulating levels of 5-HT as well as luteinizing hormone (LH) and follicle stimulating hormone (FSH). The symptoms of the disease include a reduced libido and impotence in addition to diarrhoea as a direct result of the stimulating effects of 5-HT on the intestinal muscles. In trials with PCPA in carcinoid patients[22] the drug was effective in reducing the rate of 5-HT production by the tumour but had little effect on LH or FSH levels. Although it did relieve the diarrhoea it did not restore the potency or libido of the patients.

The endocrinological effects of PCPA in normal subjects at doses which do not cause excessive side-effects (1.5 g per day or less) seem to be minimal. After 12 days' treatment with a daily dose of 1.5 g of PCPA, six healthy male subjects showed no change in blood levels of LH, FSH, growth hormone, thyroid stimulating hormone, or testosterone.[23] No changes in sleep pattern or sexual behaviour were experienced by any of the subjects, although they experienced the normal incidence of the other side-effects associated with PCPA. Subsequent testing of higher doses of PCPA produced equally negative results. In studies of this type it is important that the trial of the drug be well controlled and carried out on a double-blind basis. On the occasions when an attempt has been made to cure impotence with a new drug or treatment, the patient's expectations have always been an important factor.

In spite of all the disappointments with the use of PCPA in humans, particularly after the results obtained in animals, PCPA briefly became available as a street drug in California under the name of 'Steam', but it has never become a significant problem as a drug of abuse.

Drugs with Aphrodisiac Side-effects

Scattered throughout the medical literature it is possible to find numerous anecdotal case reports of patients who have experienced an increased libido during a particular course of drug treatment. Apart from the dopamine- or serotonin-related compounds already mentioned, the most common group of drugs associated with increased sexual desire are the antidepressants. The complicating factor here is that, in the absence of any controlled trials in non-depressed subjects, it is impossible to say how much of the effect observed is due to the relief of the depression and the improved state of mind of the patient induced by the drug. However, most of the drugs used in depression potentiate the actions of the adrenergic nervous system. In the periphery this can influence the neuronal control of the penile erection and ejaculation. The commonest symptom produced is delayed ejaculation which, although it can be an advantage in some cases, could lead to increased sexual problems in patients who find ejaculation postponed indefinitely.

The potential value of this side-effect of antidepressant therapy is illustrated by the comments made by a doctor[24] on the medical record form of a patient being treated with the antidepressant drug clomipramine (Anafranil):

'. . . incidentally, his wife − a big film star type girl − wants him to stay on the drug as he can maintain an erection for considerably longer than ever before which is helping to resolve one of *her* upsets'.

If ejaculation can be delayed without any loss of sexual interest or feelings of frustration at a dose that is within the safe therapeutic range, there may be a case for the use of these types of drug in the treatment of premature ejaculation. In a study reported in 1973,[25] clomipramine appeared to be highly successful in this respect, as long as the dosage was carefully adjusted to suit individual patients' needs.

The stimulatory effects of tricyclic antidepressant drugs on sexual performance are not confined to male patients; in a Japanese study of imipramine and trimipramine, five patients reported increasing sexual interest which, in three women, amounted to nymphomania. One male schizophrenic had uncontrollable urges to masturbate in public. Occasionally, clomipramine has produced difficulties in achieving orgasm in women, but it has also been tried as a treatment for frigidity. The end result is clearly very dependent upon the mental state of the individual patient, and it is impossible to predict with certainty whether the drug will have a beneficial or detrimental effect on sexual function. This is hardly the ideal clinical situation.

A different class of antidepressent drug, chemically unrelated to those already mentioned, is represented by nomifensine (Merital). A recent report from Australia[26] describes improved sexual function in one male and one female patient. It is known that nomifensine can inhibit prolactin release by virtue of its dopamine-like properties, so it is possible that this particular drug is acting by the same mechanism as *L*-DOPA and bromocriptine to increase libido.

Some of the drugs used to treat hypertension have also been associated with increased sexual appetite, in particular hydralazine, prazosin and labetalol, but there are also numerous instances of the same drugs causing diminished libido. It is not likely that the restoration of a normal blood pressure in itself will affect the cerebral aspects of the sex drive, although the vasodilatation produced by the drugs could affect the ability to achieve and maintain an erection.

Another recent study carried out in Australia[27] on ten normal healthy male volunteers with a relatively new antihypertensive drug, endralazine, found that intravenous infusion of the drug caused penile erections in four of the subjects, feelings of sexual arousal but no physical manifestations in a further three subjects, and no effects at all in the remainder. The time course of the effects coincided with the period when the blood pressure was lowered. Although it is interesting that sexual arousal could be produced without an erection occurring, it is unlikely that endralazine will ever find use as an aphrodisiac.

Drugs with Anaphrodisiac Side-effects

The detrimental effects of drugs on sexual performance are almost invariably observed in male subjects. This is because the male requires a fully functioning peripheral nervous system in order to achieve erection and ejaculation, and these are essential to a satisfactory sexual act. It also means that there are many more potential sites at which a drug can act to interfere with sexual performance. In the female, there is still some doubt concerning the physiological nature of the orgasm and, in any case, few women seem to expect an orgasm every time they have sex.

The drug effects which can lead to sexual dysfunction are set out in Table 8.4.

There are general classes of drug which, by the nature of their pharmacological properties, will tend to have detrimental effects either on the libido or on the erectile capacity of the patient. The major tranquillizers or antipsychotics are all associated with reduced libido and a low but consistent incidence of male impotence. Sedative-hypnotic drugs at moderate to high doses will also reduce sexual performance in the male. Some of the drugs already mentioned as possibly being aphrodisiac are often anaphrodisiac at higher doses. The drugs that have been implicated in disorders of ejaculation include thioridazine, chlorpromazine, chlorprothixene, chlordirazepoxide (Librium), perphenazine, trifluoroperazine, butaperazine, reserpine, haloperidol, guanethidine and phenoxybenzamine.

Erectile impotence has been reported with chlorpromazine at higher doses, imipramine, protriptyline, desipramine and mebanazine.[24] Over a large number of patients it is possible to indicate the likely incidence of impotence, for example:

Drug	Frequency of impotence (%)
Isocarboxazid	1
Phenelzine	2
Iproniazid	3
Imipramine	2

The phenothiazine and butyrophenone classes of antipsychotic drugs have anticholinergic, anti-adrenergic, and antidopaminergic activity. These actions will directly affect both the erectile response and the libido. Antidopaminergic drugs tend to cause the blood levels of prolactin to rise, and this reduces the libido. Unfortunately, these properties are essential for the therapeutic action of the drugs, so it appears that some side-effects are unavoidable. Among the phenothiazines, thioridazine is notable for its peculiar ability to induce aspermia

Table 8.4: Drug Effects Producing Sexual Dysfunction

Male Nature of Dysfunction	Cause	Female Nature of Dysfunction	Cause
(1) Decreased libido	Dopamine receptor blockade in the hypothalamus, anti-androgenic actions, general depressant actions	(1) Decreased libido	As in male patients
(2) Diminished erection	Increased activity of sympathetic nerves, decreased peripheral blood flow	(2) Reduced vaginal lubrication	Anti-cholinergic actions, inhibition of normal secretory processes
(3) Decreased ejaculation	α-Adrenergic receptor blockade	(3) Delayed or inhibited orgasm	Reduced peripheral blood flow, local anaesthesia
(4) Depressed erection and ejaculation	Reduced spinal reflexes, blockade of ganglionic transmission		
(5) Delayed or inhibited orgasm	Excessive sympathetic nerve activity, local anaesthesia of genital areas		

(no ejaculate after an orgasm), with an incidence of up to 30 per cent of patients. It seems that the ejaculate is produced by the various accessory glands in a normal way, but that it is then forced backwards into the bladder instead of outwards along the urethra. This is possibly due to alterations in local muscle tone resulting from blockade of α-adrenergic receptors.

Several diuretic drugs have been reported to produce impotence in up to 36 per cent of male patients. With the thiazide diuretics the mechanism of action on the erectile response is not known, but in the case of spironolactone, which produces impotence and loss of libido in both sexes, it would seem that the drug has significant anti-androgenic activity. This means that it will compete with the circulating male sex hormones. A thorough review of the drugs which can impair sexual functioning has been compiled by Buffum (1982), who also provides a full bibliography of original papers in which the effects have been reported.

Anti-androgens

The anti-androgens have found use in the treatment of habitual sex offenders who are suffering from an excessive libido. They are also used to treat carcinoma of the prostate, and severe hirsutism in women. The most potent and widely used drug is cyproterone acetate (Androcur). This inhibits sperm production and induces reversible infertility. Sexual arousal does not disappear completely and sexual preferences are not apparently altered. Progestins have also been tested in sex offenders, the object being to reduce the libido without causing complete impotence. However, it is only in cases where the offences are of a habitually violent or deviant nature that this type of drug treatment is used.

9 APHRODISIACS IN THE FUTURE

Introduction

So far in this book we have considered the use of aphrodisiacs from about 3000 BC to the present day. In this final chapter we shall be examining the possible aphrodisiacs of the future. It is most unlikely that a pharmaceutical company would undertake the necessary research and development required to produce a drug which could be marketed as an aphrodisiac. It has been estimated that each successful new drug introduced on to the market represents at least seven years' development work and an investment of between ten and twenty million pounds. Add to this the problems of assessing aphrodisiac activity, and the need to convince the Committee on the Safety of Medicines that there is an ethical clinical role for such a drug, and the likely return on the investment becomes a doubtful economic proposition. On the other hand, there is always the possibility that a newly introduced drug turns out to have significant aphrodisiac side-effects. It is very much easier to obtain approval for a novel use of an existing drug which has already undergone all the necessary toxicological and clinical screening than to satisfy the safety requirements for a completely new compound.

Some drugs have already been used in the treatment of sexual dysfunction, notably testosterone and LHRH, but these are only of value in cases where the libido or performance is diminished as a consequence of a disease state; there is no evidence that either hormone will act to increase the sexual drive above normal levels. LHRH or one of its synthetic analogues is arguably the most exciting prospect for patients with pituitary or hypothalamic disorders of brain function. The illicit-drug-using community will no doubt be swift to try any new drug with potential aphrodisiac side-effects but, as a population group, they are notoriously poor at discriminating between real and imagined drug effects. Their opinions and preferences are of dubious value.

Alternatives to drugs are available for provoking or increasing sexual desire, and the possibility of exploiting a natural human sex pheromone should not be discounted. In fact, there is already some evidence that our relatively neglected olfactory sense plays an important role in sexual attraction. The very size and prosperity of the perfume industry testifies to the fact that we are sensitive to, and appreciative of, exotic odours.

One form of alternative medicine that has recently attracted attention is aromatherapy, the use of fragrant oils to obtain therapeutic responses. The association of odours with particular experiences or emotions can lead to a form of conditioning so that the odour itself eventually becomes sufficient to influence the mood. With this potential application of classical conditioning processes we have returned full circle to the basic idea presented in the first chapter; that is, the most effective aphrodisiacs work through the mind rather than the body.

Pheromones

Pheromones are a special type of hormone which act externally and are a means by which animals can communicate over remarkably long distances. They can be used to give out warnings (the smell of fear), to discourage attacks from predators (the offensive odour of the skunk), or to attract members of the opposite sex (the sex pheromones). It is this last group which can perhaps be considered as the natural aphrodisiacs. Sex pheromones have been positively identified in bacteria, fungi, fish, molluscs, insects, reptiles, birds and mammals, so it would seem that the aphrodisiac is a device used by a considerable proportion of the animal kingdom. On this basis it is logical to assume that the human animal is no exception to the rule and that we do possess our own natural sex attractants.

Pheromones can act in two basic ways; first, to directly alter the observed behaviour of other animals (the releasing effect), and secondly, to produce changes in the hormonal balance of the target animals (the priming effect). This last action does not necessarily produce an obvious change in behaviour and is consequently more difficult to identify. The intricate social structures of bee and ant colonies are almost entirely controlled by pheromones produced by the queen in the colony. The swarming of locusts is also thought to be initiated by the release of pheromones.

Many sex pheromones have relatively simple chemical structures; some examples from a variety of species are shown in Figure 9.1. Gyplure is the sex attractant released by the female gypsy moth, and some idea of its potency can be gained by considering that a single moth contains only $0.01\,\mu g$ of gyplure and yet, by secreting this pheromone, she can attract male moths from distances of over

Figure 9.1: The Chemical Structures of Some Pheromone Sex
Attractants

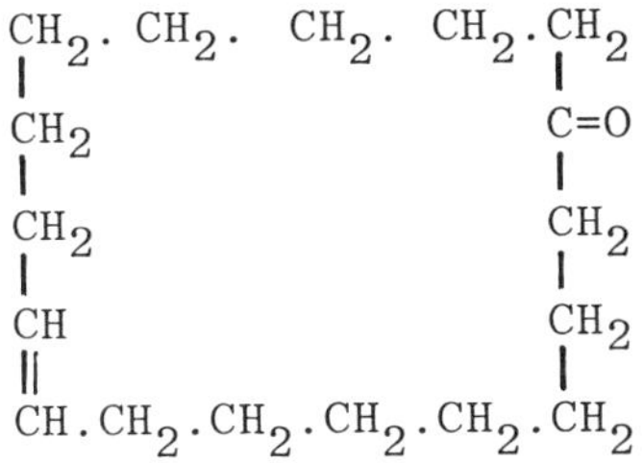

Civetone

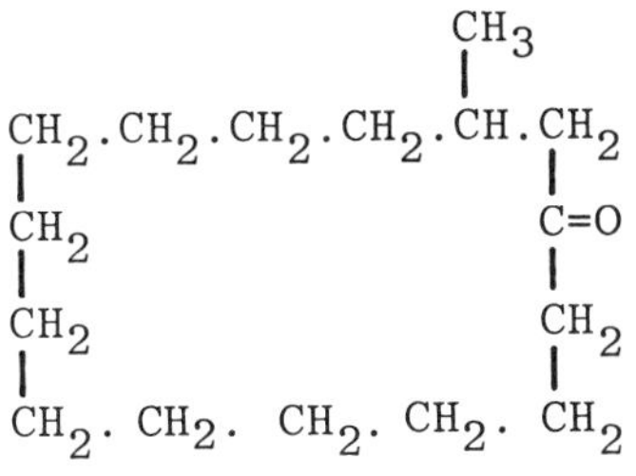

Muskone

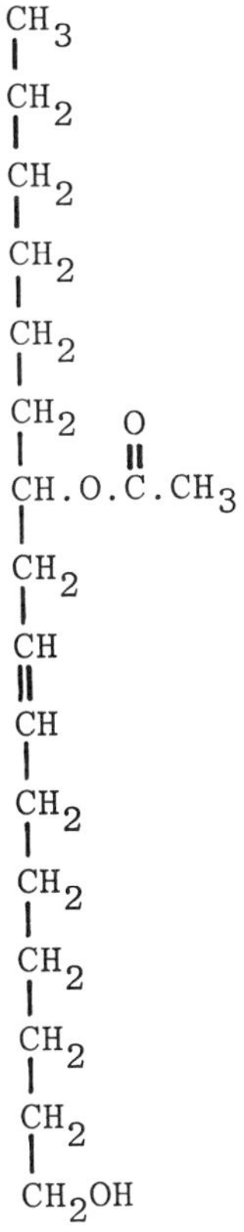

Gyplure

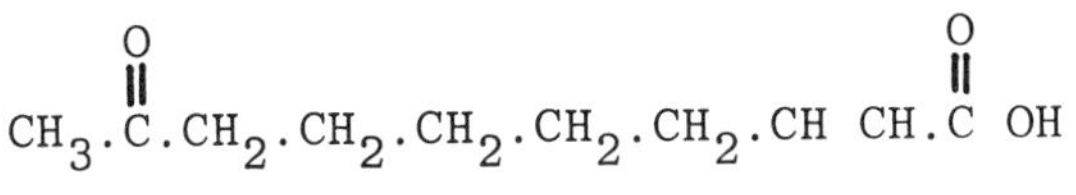

Queen Bee Substance

3 km downwind. From these figures it has been estimated that the
threshold concentration of gyplure required to have a positive effect
on the male is only 200 molecules per cubic centimetre of air. Civetone
and muskone are secreted by the civet cat and musk deer, respectively;
they are thought to be specific sex attractants, although their exact

mechanism of action has not yet been elucidated. The chemical nature of the aliphatic organic acids or 'copulins' has been determined from experiments with rhesus monkeys (see below).

Insects are relatively easy to study since most of their behaviour is instinctive and can be categorized into clear stereotypes. Mammalian studies are far more difficult to conduct and there is relatively little direct evidence of the action of sex pheromones in higher animals. In male mammals the secretion of pheromones may be associated with the marking out of territory, the attracting of females, and possibly the stimulation of oestrous in females (i.e. a priming action). Obviously, in the wild, the only profitable time for a male and a female to meet is when the female is on heat and able to conceive, a situation which would be facilitated by the release of a priming pheromone. Dr Alex Comfort has stated that man still has the necessary secretory glands for releasing pheromones, and that if he does not use them, then he is probably the only mammal that does not. It is unlikely that we have lost through evolution all our ability to secrete sex pheromones; the time scale is too short for one thing, but it is far more likely that, through social pressures, we are gradually losing the ability to detect and to respond to pheromones.

Humans are unlike most other mammals in that there is no time during the menstrual cycle when the female is not able to enjoy sexual intercourse. There is therefore little biological need for a priming sex pheromone, although it is not impossible that some natural secretion can act at the cerebral level to stimulate an interest in sex. However, in humans there are marked sex differences in the ability to smell and taste certain compounds; for example, exaltolide (a lactone of 14-hydroxytetradecanoic acid) can only be clearly detected by adult females prior to the menopause, and then most strongly at the time of ovulation or oestrous. Males are completely unable to detect exaltolide, although some experiments have been carried out in which males, injected with the female hormone oestrogen, became capable of detecting the chemical.

Perfumes

If the discussion so far has appeared to have little relevance to the subject of sexual attraction in contemporary society, it should be noted that muskone, civetone and exaltolide are all used as ingredients in perfumes by the cosmetics industry. Manufacturers often adopt a more or less overt sex attractant theme in their marketing and advertising; one range of products at the lower end of the market is actually called 'Aphrodisia'. From previous consideration of the motives for buying a purported aphrodisiac it is clear that the company in question would be better off reserving that particular brand name for their most expensive product.

Perfumes have always been valuable for increasing sexual attractiveness in both sexes as well as for the more obvious purpose of overcoming less pleasant odours. The Arabs were great exponents of the art of perfumery, having discovered the techniques for distilling the essential oils from plants, and it is surely no coincidence that one of the best known erotic Arab texts is entitled *The Perfumed Garden*. The author of this book explains how the use of perfumes can assist a man in getting possession of a woman, and he relates the story of Mocailama and the prophetess Chedja. Mocailama is given the following advice for achieving the seduction of the prophetess:

'Fill the tent . . . with a variety of different perfumes, amber, musk, and all sorts of scents, as rose, orange flowers, jonquils, jessamine, hyacinth, carnation and other plants. This done, have placed there several gold censers filled with green aloes, ambergris, nedde and so on. Then fix the hangings so that nothing of these perfumes can escape out of the tent. Then, when you find the vapour strong enough to impregnate water, sit down on your throne, and send for the prophetess to come and see you in the tent, where she will be alone with you. When you are thus together there, and she inhales the perfumes, she will delight in the same, all her bones will be relaxed in a soft repose, and finally she will be swooning. When you see her thus far gone, ask her to grant you her favours; she will not hesitate to accord them.'

The plan succeeds, thus indicating the value of providing suitably pleasant stimuli to the senses: soft lights, relaxing music, subtle perfumes; all will conspire to encourage the appropriate state of mind.

Over the last few decades the use of deodorants and antiperspirants has become almost universal in North America and Western Europe, and they now represent a major industrial sector. The function of these products is to mask the natural 'unpleasant' odour of the human body and replace it with a partially synthetic 'pleasant' odour in combination with extracts of the glandular secretions of lower animals.

Perhaps as a consequence of the development of civilization some of our senses have become dulled through disuse, but there still remains an important olfactory component in sexual attraction. If we are to believe the perfume advertisers, a woman needs only to apply some of their (usually expensive) product to strategic areas of her body and she will immediately attract the attention of every male within range. However, the girls who feature in advertisements of this type are always strikingly beautiful and hardly in need of any additional chemical aids to gain the attention of the opposite sex. On the other hand, it would seem highly inappropriate for an elderly woman to use a strong perfume which would be sending out a misleading social signal. The best perfumes are perhaps those which do not commit olfactory assault but send out a more subtle message of sexual availability. The average British male is not yet in the habit of using perfumes, although most deodorants contain some scented ingredients and would be more accurately described as reodorants. Future research may make it possible for the cosmetics companies to add compounds of established potency as sex pheromones to commercially available products; indeed, this may already have happened.

Aromatherapy

Only a few people are fortunate enough to possess a total olfactory memory, and they probably find ready employment as wine or tea tasters, but most people are able to recognize a familiar perfume, perhaps on a piece of clothing, and can immediately recall the particular person with whom it is associated. This association, or conditioning, can be exploited in several ways; in a commercial context it can be used to market a perfume, and in the form of alternative medicine known as aromatherapy it can be used to achieve a therapeutic response. The man who splashes himself with a well-known make of after-shave surely realizes that it will not make him look like a world champion racing driver, or even drive like one, but the readily purchased ability to *smell* like a world champion is sufficient in itself to persuade the customer to buy the product. This is a very simple example of identification by association, but aromatherapy operates by determining an individual's personal responses to particular aromas.

If an odour is paired with a particular emotional state, it becomes theoretically possible to induce that state by presenting the subject with the associated odour. There is no doubt that specific fragrances can have profound psychological effects. Olfactory signals enter the brain at an older and more primitive region compared with the visual and auditory inputs which are closely linked to the higher centres in the cortex. This difference reflects the evolutionary antiquity of smell and taste as sensory functions, and means that odours can act at a subconscious level to evoke memories or moods.

The technique in aromatherapy is basically simple. Taking as an example the treatment of anxiety, the patient is encouraged to relax by thinking of some pleasant memory such as a seaside holiday. This is reinforced by introducing an essential fragrance associated with the seaside, the chemical known as maritima, for example. While in this reflective state the patient is gently encouraged to face their particular anxiety-inducing factors. These are incompatible with the relaxed state produced by the olfactory recall and so the patient becomes less sensitive to his previous fears or anxieties. The advantage of using fragrances is that the patient can use the technique on his own to achieve the relaxation that would normally require the presence of the psychiatrist.

Although relaxation therapy and the treatment of phobias are seen at present as the main applications of aromatherapy, it takes little imagination to appreciate how, in the future, this might be applied to helping in cases of sexual dysfunction. The association of past sexual pleasure with a specific fragrance has latent possibilities.

Bodily Secretions

If artificial perfumes can act as psychological sex stimulants by association, do the natural human body odours have any positive effect on sexual arousal? A large proportion of the unpleasantness in body odour is not due to the glandular secretions themselves but to the actions of bacteria on the surface of the skin converting the components of the secretions into more odoriferous products. The cheesy odour of feet, for example, occurs only in people who wear shoes and socks in a warm climate. The microclimate of warmth and high humidity within the shoe makes it an ideal breeding ground for bacterial and fungal colonies.

Of all the various glandular secretions of the body it is those produced by the vagina that are most likely to contain hormones or pheromones capable of influencing the behaviour of the male. It is known that male rhesus monkeys are only attracted to the females during oestrous, and that the application of vaginal secretions from females in oestrous to monkeys who are not ovulating will cause the males to take an immediate interest and attempt coitus where previously they had remained indifferent.[1] This suggests that the vaginal secretions contain some active principal which stimulates the males into this behaviour pattern. These unknown factors were named 'copulins', and it was found that they could be extracted from the secreted material by fairly simple chemical procedures. The principal constituents of these extracts were found to be a series of aliphatic organic acids. An artificial mixture of these acids, made up in the correct relative concentrations, was then tested and found to produce the same mounting activity and copulatory behaviour in the male monkeys as the authentic vaginal secretion. The formula of this synthetic mixture (in terms of the concentration in micrograms per millilitre of ether extract) was as follows:

acetic acid	9.2
propionic acid	8.8
iso-butyric acid	4.2
n-butyric acid	12.8
isovaleric acid	8.3

The results in the rhesus monkeys suggested that these very simple organic compounds were the copulin factors in vaginal secretions. The next step was to conduct similar experiments in human volunteers.

A rather limited study was carried out involving the chemical analysis of the vaginal secretions from three pre-menopausal and two post-menopausal women.[2] The secretions from all subjects showed a similar pattern of acetic, propionic, valeric and caproic acids, but the concentration of *n*-butyric acid was 50 times higher in the three pre-menopausal women. This might imply that *n*-butyric acid is linked in some way to normal sexual functioning in the female and, following this discovery, an attempt was made to investigate the sexually stimulating effects of a synthetic mixture of these aliphatic acids in human subjects.[3] The experiment was designed as a double-blind placebo crossover in which the aliphatic acid cocktail was tested against an alcohol–water mixture with or without added perfume. The mixtures were used in the same way as a perfume by a group of married female volunteers, and their sexual activity with their husbands was recorded over the subsequent three months by means of a questionnaire.

No differences were detected between any of the four groups of women, either in terms of frequency of coitus or the urge to indulge in sex on the part of the husbands. Although this negative result may seem at first to be disappointing, there are several possible reasons why no positive effects were seen. The conditioning of the husbands by previous exposure to a favourite perfume, or the variability in the olfactory sensitivity of the husbands (smokers, for example, generally have a poorer sense of smell than non-smokers), may have been responsible for the absence of any effect, or possibly the synthetic mixture did not correspond closely enough with the natural vaginal secretion or did not contain some other unsuspected vital ingredient. The original recipe for the mixture was, after all, based on data from only five subjects: they may not have been representative of the population as a whole.

Children have been shown to possess completely different preferences for smells as compared with adults. They tend to favour fruity odours where adults favour flowery odours; indeed, some popular perfumes have been ranked as 'quite revolting' by children.[4] The changes in patterns of preference occur immediately before and during puberty, so it is possible that the alterations in hormonal balance which occur at this time have a direct effect on the olfactory sense.

There has been very little research into sex differences in olfactory sensitivity, although the food and cosmetics industries do carry out extensive investigations into consumer preferences for odours or flavours. Some of the most interesting work on sex differences in the response to hormones and related chemicals has been carried out using androstenol. The initial results proved sufficiently promising for one enterprising perfume manufacturer to begin marketing 'Andron', a perfume whose main ingredient was androstenol.

Androstenol is a steroid found in the urine of both adult men and women as well as the axillary (armpit) sweat of adult men. The early experiments consisted of making the subjects, of either sex, wear face masks impregnated with androstenol or a control substance, and then asking them to assess photographs of members of the opposite sex. The number of favourable judgements was significantly increased when the mask was treated with androstenol compared with controls. However, this experiment represents an extremely artificial situation, in which the postulated pheromone is not really being used to communicate between individuals, and is more akin to the simple conditioning situations described earlier.

A better-designed experiment has recently been published,[5] although the results are not likely to encourage any other perfumiers to enter the pheromone market. The subjects in this case (39

Table 9.1: Effects of Exaltolide and Androstenol on Perceived Physical Attractiveness

Condition	Male Confederate		Female Confederate	
	Number	Rating	Number	Rating
No odour	13	7.85 + 1.28	12	6.84 + 1.40
Exaltolide	13	7.54 + 0.66	12	7.08 + 0.90
Androstenol	9	7.11 + 1.05	12	6.17 + 1.47

undergraduate students of either sex), were assigned to androstenol, a synthetic musk (exaltolide), or control treatments. The substance was applied in the same way as a perfume, and no other perfumes, deodorants, soaps, shampoos, etc. were permitted. The solution to be tested was applied to the skin of a 'confederate' subject who was then introduced to the member of the opposite sex, the 'naive' subject. During the test period both subjects were asked to view a set of innocuous slides and rate their pleasantness. The impression was given that this was the object of the experiment. There was, however, a critical question concealed within the questionnaire in which the subjects had to rate the physical attractiveness of their partner on a scale of one to ten. After the test each subject was given samples of the test substances on filter paper to see whether or not they could smell them. The data from those who could not detect androstenol were excluded from the analysis.

The results are shown in Table 9.1, and it is immediately obvious that neither the androstenol nor the exaltolide has made any difference in the assessment of physical attractiveness. When the results were subjected to statistical analysis, the only significant conclusion that could be drawn was that the male confederates were, on average, judged to be more attractive by the females than were the female confederates by the males. Even this experiment has severe limitations in its design; people vary widely in their sense of smell, and there is no real reason why one should expect a pheromone to be detectable at the conscious level. The dose of androstenol may have been below the threshold of effect and, in the case of the female students, it would have been useful to eliminate those taking a contraceptive pill and to test the others at the same point in their menstrual cycle. Scientists are always very ready to think of reasons why an experiment gave a negative result, and much less ready to question a positive result which supports their original hypothesis. In this case there is every prospect that there will be many more investigations into potential human pheromones in the future.

Brain Stimulation

The principal disadvantage of drugs that act within the brain is that they tend to act at a number of sites throughout the central nervous system, so that their effects are less specific than one would wish. For example, the drugs used as antipsychotics work by antagonizing the effects of the neurotransmitter, dopamine. This action is essential to their effectiveness in treating diseases such as schizophrenia. These drugs work very well to relieve the symptoms of psychotic illness, but there are obviously other sites in the brain where dopamine has an entirely different function, so that certain side-effects become inevitable with the use of these drugs. One particular problem is the condition known as Parkinsonism, which can be overcome by using drugs which mimic the actions of dopamine (see Chapter 8).

An alternative means of making the effects of drugs more specific is to target them at the specific site where they are needed. It is relatively easy to achieve this experimentally by inserting a cannula into the brain, under general anaesthesia, and then applying drugs through the cannula, using antiseptic conditions, directly into the region being investigated. This, however, is hardly a practical scheme in human subjects.

Another method of investigating specific regions of the brain involves the insertion of fine electrodes which are insulated except at the extreme tip. Because of their small size they cause negligible damage to the surrounding brain areas, and they can carry a small electrical current which is capable of stimulating the region of the brain immediately in the vicinity of their tip. The early experiments of this type were carried out in animals, and by stimulating different areas of the brain it was possible to build up a map of the functions of the various brain regions.

In the septal areas of the rat brain, which lie deep below the cortex, it was found that mild electrical stimulation could produce either rage or fear, depending upon precisely where the electrode was placed. In a more sophisticated series of experiments the animal was able to operate a lever which controlled the rate of electrical stimulation it received. At certain electrode positions in the brain the animal would spend the majority of its time in self-stimulation, using the lever at the expense of time normally spent eating and drinking. From the results obtained in experiments of this type it has been concluded that there exist, in the brain, very specific pleasure centres which are amenable to artificial stimulation resulting in intense feelings of pleasure.

Although these experiments may not immediately seem to have many practical applications to the human condition, intra-cranial self-stimulation, as it is called, has proved very useful in the treatment of previously intractable epilepsy. An epileptic fit arises as a local electrical disturbance or instability of the neuronal membranes at a discrete point, or focus, in the brain. If the position of this focus can be identified, and a stimulating electrode placed at a site which has an inhibitory influence on the nerve cells in the focus, the electrical disturbance can be prevented from spreading to the surrounding brain areas and producing a full-blown epileptic fit. This technique has been tried successfully in several patients who have not responded to the normal regime of anticonvulsant drug therapy. It was mentioned earlier that some types of epileptic fit were occasionally preceded by orgiastic feelings of pleasure. In these rare cases, the focus of the electrical disturbance is likely to be in the region of the septal pleasure centres.

The Tulane University School of Medicine in the United States has for many years been conducting experiments using direct brain stimulation in conscious human subjects.[6] The purpose of most of these experiments was to improve the mental state of severely withdrawn schizophrenics who were exhibiting total apathy, listlessness and lack of interest in their surroundings. As with the earlier animal studies, it was quickly found that the septal region of the brain was the most promising area to stimulate in order to elicit pleasurable sensations. The patients became much brighter, looked more alert, and seemed more interested in and attentive to their surroundings.

The electrodes were implanted by surgical procedures and caused no discomfort when the patients recovered from the anaesthetic. The patients were not told when the stimulating current was being applied so that their verbal responses to questions concerning their mood and feelings would be unprejudiced. In one instance a young man had been on the verge of tears as he described his father's near fatal illness, and was condemning himself for being somehow responsible. As the septal stimulation was begun, his conversation immediately turned to plans to make a date and seduce a girlfriend. When he was asked afterwards he was totally unable to explain why his mood and conversation had changed so suddenly. This phenomenon was repeated several times.

Other patients suffering from severe pain as a consequence of advanced carcinomas also experienced pleasure and relief from the intense pain and anguish from which they were suffering. Very occasionally, overt symptoms of sexual arousal occurred and some non-schizophrenic patients experienced erections during the period of stimulation. The pleasure sensations evoked from the septal region very commonly have a sexual component, and this suggests that the cerebral aspects of sexual pleasure and arousal may be related to this region of the brain.

In some self-stimulation experiments in which the patients had a choice of three switches controlling three different electrodes, only one of which was placed in the septal area, the patients very soon devoted their attention exclusively to the switch operating the septal electrode, after first randomly exploring the effects of each of the three switches. One subject, on being asked why he was switching on the stimulation so frequently, replied that it made him feel good and as though he was building up to a sexual climax. He did not however achieve an orgasm by this means.

Although these experiments may seem to be only one step beyond physical self-stimulation, a sort of cerebral masturbation in fact, they do indicate the potential scope for the employment of intra-cranial self-stimulation in the treatment of frigidity or psychological impotence. At present, the necessary surgical procedures are not without risk and also raise several ethical questions. The idea of obtaining an electrode implant within the United Kingdom National Health Service is not a likely proposition in the foreseeable future. But at a time when private medicine is prepared to offer, and undertake, extensive surgery for sex-change operations, this neurophysiological solution to some psychosexual problems is perhaps not too far away.

Subliminal Stimulation

It has been known for some time that it is possible to convey simple visual messages by showing them on a film or video screen for very short periods of time, normally between 25 and 40 ms. The signal is so brief that the subject is not consciously aware of having seen it, but the brain apparently does take notice of the signal, which can then have a subtle suggestive influence on the subject. Not all psychologists are convinced that subliminal signals are effective in altering behaviour, but there has been enough evidence available to persuade the authorities to make subliminal advertising illegal.

The technique has also been used by the makers of horror and suspense films: by flashing words such as 'blood' or 'death' on to a corner of the screen for 30 ms or so, it was hoped that the impact of the film could be increased as a result of the subliminal cues delivered to the audience. The potential of this method for increasing sexual awareness rather than fear should be immediately apparent. An American company has recently introduced a device called Expandovision which enables the user to add subliminal messages to his television screen. Whether a message such as 'Let's go to bed' would have the same seductive impact as soft lights, sweet music, and a moderate dose of champagne is open to question, but only time will show whether Expandovision is the first of a new generation of electronic aphrodisiacs.

Conclusion

From the evidence of the last 5000 years there can be little doubt that there will always be a future for aphrodisiacs. Whether or not they actually work is probably of no consequence: people will always believe what they want to believe. Advances in medical science may yield drugs or other treatments which can dramatically improve the prospects of patients with specific sexual dysfunctions, but for the healthy normal subject who seeks only to add new dimensions of pleasure to his or her sex life, the quacks will be ever ready with their nostrums and remedies for separating the credulous from their money. In spite of our ever-increasing level of general education and the spread of scientific knowledge among the population, the suckers still seem to be born at the traditional rate. It is perhaps appropriate to end with a line by the poet William Blake:

'Everything possible to be believed is an image of truth.'

APPENDIX 1: LIST OF APHRODISIACS

Throughout this book, vast numbers of animal, vegetable and mineral substances have been considered as potential aphrodisiacs on the basis of their historical or mythical reputations. Although the large majority of purported aphrodisiacs can be immediately refuted as totally spurious, the enduring credulity of mankind is amply illustrated by the following list of the potential aphrodisiacs referred to in the principal references quoted in the text. The sheer number and variety of substances should be sufficient to induce a healthy scepticism in the reader:

Absinthe
Acorus calamus (sweet flag)
Adrenaline
Adrenochrome
Advocaat
Affion (Chinese opium)
Agate (precious stone)
Alchone (herb)
Alcohol (in all forms)
Allspice
Almonds
Almond soup
Aloes (bitter essence)
Alpine gentian
Amanita muscaria (toxic fungus containing muscarine)
Amaranth (love-lies-bleeding)
Ambergris
Anacyclus pyrethrum
Analeptics (stimulant drugs)
Anchovy
Angel water (an early perfume)
Anise (source of aniseed)
Ants (substitute for cantharides)
Anvalli (tropical nut)
'Aphrodisin' (proprietary remedy containing yohimbine)
Apium petroselinum (parsley)
Apple
Apricot brandy
Aquamarine (precious stone)
Areca palm (betel-nut tree)
Armagnac
Arrack

Arris root
Arrowroot
Artemisia (wormwood)
Artichoke (globe)
Asparagus
Aubergine
Avocado pear
Bamboo shoots
Banana
Banisteria caapi (plant source of harmine alkaloids)
Banyan
Barbel
Basil (sweet herb)
Bathing
Beans
Beef
Beer
Beetroot
Belladonna (deadly nightshade, source of atropine)
Benedictine
Betel nut
Bhang (cannabis)
Bhuya-Kokali (plant of the Solanaceae family; contains alkaloids)
Bird's-nest soup
Birthwort
Blood
Bois Bande (West Indian tree)
Bones (of peacock or hyena)
Borax
Brains

Brandy
Brassica eruca (rocket)
Brewer's yeast
Broad-bean soup
Broccoli
Bufotenin
Burgundy wine
Buttermilk baths
B-vitamins
Cabbage
Cactus (hallucinogenic *peyote*)
Calabash (gourd)
Calamint
Calcium
Calisia (Peruvian tree bark)
Camel bone
Camel's fat
Camel's milk
Camphor
Cannabis
Cannelloni
Cantharides
Caperberry
Capsicum (pepper plant)
Caraway
Cardamon
Cardoon (artichoke-like plant)
Carrot
Castor oil
Caviar
Celaton CH3 plus (procaine)
Cevadille (spurge plant)
Celery
Champagne
Chartreuse liqueur
Cheese
Cherries
Chervil
Chestnuts
Chick peas
Chicken
Chillies
Chives
Chocolate
Chutney
Cider
Cinchona
Cinnamon
Civet
Clams
Cloves
Cocaine
Cockle bread
Cockles

Cod liver oil
Cod's roe
Cointreau
Cola drinks
Cola nuts
Colewort
Coriander
Cow wheat
Crab
Crab apples
Crayfish
Cream
Creme de menthe
Cress
Crocodile tail
Crocodile teeth
Cubeb (pepper berries)
Cucumber
Cumin
Curry
Cuttle-fish
Cyclamen root
Damiana (plant extracts)
Dandelion wine
Darnel (a species of grass)
Dates
Deer antlers
Deer sperm
Diasatyrion (plant root)
Dill
Dog-stones (orchid root)
Dove brains
Dragon's blood (plant resin)
Drambuie
Drepang (sea slug)
Dried frog
Dried liver
Duck (Peking)
Dufz (perfume)
Dumplings
Eels
Eggs
Egg-plant (aubergine)
Elderberry wine
Endive
Eryngo
Euphorbium
E-vitamin
Falernia
Fennel
Figs
Fish
Fleawort
Frangipane (*Olibanum*)

French Onion Soup
Frogs' bones
Frogs' legs
Fruit
Galanga (plant root)
Galantine
Gall
Game
Garlic
Gentian
Ghee
Gillyflower
Ginger
Ginseng
Glucose
Goats' eyes
Goat suet
Goats' testicles
Goose
Gopalika (plant)
Gossypion
Goulash
Grape juice
Green pepper
Guduchi (Indian plant)
Guinea fowl
Haddock
Haggis
Halibut
Hallucinogenic drugs
Ham
Hare soup
Haricot beans
Harmine (alkaloid)
Hashish
Hedysarum gangeticum
Hellebore
Hemlock
Henbane
Henna
Herbs
Herissah (dish of mutton and pepper)
Herring
Hippomanes
Honey
Horseradish
Hydrocotyle
Hydromel (honey and water)
Hyssop
Ibogene
Indian hemp
Intestines
Iodine
Iron

Isinglass
Italian soup (calves' heel, crayfish, carrots, celery and shallots)
Jasmine
Julep
Julienne (soup)
Juniper
Kahawi (fish)
Karengro (plant)
Kasurika (Indian fruit)
Kava
Kedgeree
Keitafo banlon (contemporary quack remedy from Hong Kong)
Kelp
Kidney
Kidney bean
Kipper
Kite (dead flesh of)
Kshirika (plant)
Kuili (Hindu remedy)
Kummel
Kyphi (Egyptian remedy)
Lamb
Lamprey
Lard
Lavender
Lecithin
Lentils
Lion fat
Liquorice
Liver
Lizard
Lobster
Lotus flower
Lycopodium (root)
Mackerel
Madeira wine
Maerua arenaria (Indian herb)
Ma-Fu-Shuan (Chinese remedy)
Maidenhair fern
Maize
Mallow
Mandrake
Mango
Marijuana
Marjoram
Marrow (vegetable)
Marzipan
Mastic resin
Mead
Meadow sweet (*Filipendula ulmaria*)
Meat
Melon

Membrum virile
Mescaline
Milk
Mineral waters
Mint
Moh tree (source of arrack)
Molasses
Molluscs
Moly (wild garlic)
Mugwort
Mulberry
Muscat (wine)
Mushrooms
Musk
Mussels
Mustard
Mutton
Myosotis (forget-me-not)
Myrrh
Myrtle
Nail parings
Navelwort
Necks of snails
Nectar
Nectarine
Nedde (Arab perfume)
Negus (mulled wine)
Nepenthe (opium)
Nettles
Niacin
Nicotine
Ninjin (Japanese plant root)
Nuoc-Man (Chinese remedy including
    decayed fish and salt)
Nutmeg
Nuts
Nux vomica (strychnine)
Nymphaea (water lily: Hindu
    remedy)
Oatmeal
Octopus
Oils
Ointments
Olibanum
Ololuigui
Onion
Onion seed
Opium
Orchid
Origanum (marjoram)
Orris
Oysters
Papaw
Paprika

Parsley
Parsnip
Partes genitales
Partridge
Pastries
Peaches
Pears
Peas
Pellitory
Pennyroyal
Pepper
Peppermint
Perfumes
Periwinkle
Perry
Perspiration
Pheasant
Phosphorus
Pigeon
Pimento
Pine nuts
Pineapple
Pizza ururdu
Plaice
Plantain (tropical fruit)
Plovers' eggs
Plums
Polignonia (mediaeval herb)
Pomegranate
Poonac (coconut extract)
Pork in milk
Port wine
Potatoes
Poteen
Prawns
Premna spinosa (Hindu remedy)
Prune liqueurs
Pumpkin seed
Punch
Pyrethrum (pellitory)
Quail
Quassia (South American tree)
Queen apple
Quince
Quinine
Rabbit pie
Radish
Radix chinae
Rakta-Bol (Hindu name for myrrh)
Raspberries
Rauwolfia (source of resperine)
Ray
Red pepper
Reptiles

Rhubarb
Riboflavin (B-group vitamin)
Rice
Rice oil (extract of *Ruta graveolens*)
Rocket (vegetable)
Roes
Rook heart
Rosemary
Rubber
Rue
Rum
Safflower
Saffron
Sage
Salads
Salep (prepared from orchid roots)
Salmon
Samphire
Sandalwood
Scandix cerepolium
Sansevieria (plant included in a
    Hindu recipe)
Santonin (extracted from *Artemisia*)
Sapodilla
Sarsaparilla
Sassafras
Satyrion (unidentified plant used in
    Roman times)
Sausages
Sauterne wine
Savory
Scallops
Scammony
Schnapps
Sea-slug
Sedative–hypnotic drugs
Semen
Sesame
Sex glands of animals
Shallot
Sheeps' trotters
Shellfish
Sherry
Shimyaan (South African drink)
Shlakshnaparni (Indian plant)
Shrimps
Shvadaustra (Hindu recipe)
Skink
Skirret
Sloe gin
Snails
Snuff
Sole
Soups

Southernwood
Soya bean
Spanish flies
Sparrows
Spearmint
Sperm
Spikenard (aromatic extract of
    oriental plant)
Spinach
Spurges
Storgethron (Greek plant, probably
    the leek)
Stramonium (*Datura* plant)
Strawberries
Sturgeon
Surag (plant root)
Swallow's nest soup
Swans' genitals
Sweet flag
Sweet potato
Syllabub
Tallow
Taro plant
Tarpon
Tarragon
Testes
Thorn-apple (*Datura stramonium*)
Thyme
Tokay
Tomatoes
Tonka (extract from tonquin beans)
Tripe
Truffles
Turmeric
Turnips
Uchchata (Indian plant)
Udders
Unicorn
Urid seeds (Indian chick peas)
Valerian
Vanilla
Vatodbhranta (plant included in a
    Hindu recipe)
Veal
Venison
Vermicelli
Vermouth
Vinegar
Vitamins
Vodka
Vulva of the sow
Weasel ashes
Well water
Wheatgerm

Whiting
Wild poppy
Wild rue
Wine
Winged ants
Winkles
Witch-hazel
Woodcock

Woodruff
Wormwood
Yam
Yarrow
Yeast
Yellow-wood
Yoghurt
Yohimbine

APPENDIX 2: EFFECTS OF DRUGS ON THE NERVOUS SYSTEM

In order to understand how drugs exert their effects it is necessary to understand something of the functioning of the nervous system. Sexual function is dependent upon the activity of both the peripheral nervous system and the central nervous system (CNS), which consists of the brain and spinal cord. The electrical conduction of impulses along nerves provides the body with a very rapid means of communicating information. Where nerves form junctions, or synapses, a chemical neurotransmitter is released which acts like a highly local hormone and enables the impulse signal to be transmitted to the next nerve. Junctions between nerves in the periphery are called ganglia. In addition to communicating with each other, nerves, or neurones as they are termed, can transmit signals to other tissues such as the muscles or glands. Different nerves use different chemical transmitters, the best known probably being noradrenaline, the transmitter responsible for stimulating the heart and blood vessels.

Virtually all the drugs which act through the nervous system either potentiate or antagonize some aspect of the function of the neurotransmitters. Several clinically used drugs have been found to have unexpected effects on sexual functioning, usually as a result of some interaction with a neurotransmitter in an unrelated part of the nervous system. The drugs which are abused are all active in the CNS and tend to be taken precisely for the effects which would be regarded as undesirable in a clinical situation.

The main divisions of the nervous system are set out in Table A.2.1. The peripheral nervous system consists of sensory nerves which receive information from the environment, motor nerves which control the voluntary muscles, and the autonomic nerves over which we have little conscious control and which are divided into the sympathetic and parasympathetic systems. Normal sexual activity requires the intact functioning of all these different nerve types, and the CNS neurotransmitters are important in the regulation of mood, alertness and the libido.

Table A.2.1: Main Divisions of the Nervous System

System	Transmitter	Sites of Action
Cholinergic	Acetylcholine	Peripheral ganglia, motor nerves to voluntary muscles, peripheral parasympathetic nerves and throughout the CNS
Adrenergic	Noradrenaline	Peripheral sympathetic nerves and throughout the CNS
Dopaminergic	Dopamine	Nigrostriatal pathway and limbic areas of brain
Serotonergic	5-hydroxytryptamine	Striatal area of brain and some specific pathways

BIBLIOGRAPHY

This list is by no means exhaustive; I have attempted to keep the text references to a minimum while, at the same time, including some specific references to which the reader may wish to refer for further detailed information. The list below contains, for each chapter, some general sources of background information together with the specific references indicated by superscripts in the text.

Chapter 1

Chapman, H.E. *The Law Relating to the Marketing and Sale of Medicines* (Henry Burt & Son, Bedford, 1942)
Davenport, J. *Aphrodisiacs and Anaphrodisiacs* (Luxor Press, London, 1965)
Harrison, P. and Harrison, M. *Aphrodisiacs* (Jupiter Books, (London) Ltd, 1979)
Kinsey, A.C. *et al. Sexual Behaviour in the Human Female* (Saunders, Philadelphia, 1953)
Lydiate, P.W.H. *The Law Relating to the Misuse of Drugs* (Butterworths, London, 1977)
Polson, C.J. and Tattersall, R.N. *Clinical Toxicology* (English Universities Press, London, 1959)
Stark, R. (1981) *Aphrodisiacs* (Stein & Day, New York, 1981)
Wedeck, H.E. *Dictionary of Aphrodisiacs* (Peter Owen, London, 1961)
Young, J.H. *The Toadstool Millionaires* (Princeton University Press, Princeton, New Jersey, 1961)

1. Gawin, F.H. (1978) *J. Sex. Res. 14*, 107
2. Kaplan, H.S. *The New Sex Therapy* (Brunner/Mazel, New York, 1974)
3. Brown, P.S. (1977) *Med. Hist. 21*, 291

Chapter 2

The Perfumed Garden, by Shaykh Nefzawi, transl. by Sir Richard Burton (Granada Publishing, St Albans, Herts, 1963)
The Kama Sutra of Vatsyayana, transl. by Sir Richard Burton and F.F. Arbuthnot (Granada Publishing, St Albans, Herts, 1963)
'Galen's *On the Secrets of Women and on the Secrets of Men*', Levey, M. and Souryal, S.S. (1969) *Janus LV*, 208-19
Keys, J.D. *Chinese Herbs. Their Botany, Chemistry and Pharmacodynamics* (C.E. Tuttle, Rutland, Vermont, 1977)
Merck's *Manual of the Materia Medica* (Merck, Darmstadt, 1899)
Ovid (Naso Ovidius Publius), *Ars Amatoria*, transl. by B.P. Moore (Folio Society, London, 1965)
Pliny, *The Natural History*, transl. by H. Rackham (Loeb Classics Library, Heinemann, New York, 1942)

265

1. Bowers, J.Z. and Carrubba, R.W. (1978) *J. Hist. Med. 33*, 318
2. Tacquin, J. (1917) *Brit. Med. J. i*, 384

Chapter 3

Frazer, Sir J.G. *The Golden Bough*, abridged Edn (Macmillan, London, 1980)
Hughes, P. *Witchcraft* (Penguin, Harmondsworth, 1965)
Leyel, C.F. *The Magic of Herbs and Modern Book of Secrets* (Jonathan Cape, London, 1955)
McDaniel, W.B. (1948) 'The Medical and Magical Significance in Ancient Medicine of Things connected with Reproduction and its Organs', *J. Hist. Med. 3*, 525
Thompson, C.J.S. *Mysteries and Secrets of Magic* (Bodley Head, London, 1927)

1. Robertson, W.A.A. (1928) *Ann. Med. Hist. 8*, 240

Chapter 4

Acton, W. *The Function and Disorders of the Reproductive Organs in Childhood, Youth, Adult Age and Advanced Life Considered in their Physiological, Social, and Moral Relations* (Churchill, London, 1865)
Anderson, F.J. *An Illustrated History of Herbals* (Columbia University Press, New York, 1977)
British Medical Association. *Secret Remedies; What they Cost and what they Contain* (British Medical Association, London, 1909)
Culpepper, N. *The Complete Herbal*, new edition (Imperial Chemical (Pharmaceuticals), London, 1953)
Fernie, W.T. *Animal Simples Approved for Modern Uses of Cure* (John Wright, Bristol, 1899)
Fulder, S. *About Ginseng* (Thorsons, London, 1976)
Holbrook, S. *The Golden Age of Quackery* (Random House, New York, 1959)
Lucas, R. *Ginseng: The Chinese 'Wonder Root'* (Spokane, Washington, R & M Books, 1972)
Popov, I.M. and Goldwag, W.J. (1973) 'A Review of the Properties and Clinical Effects of Ginseng', *Amer. J. Chinese Med. 1*, 263
Rohde, E.S. *The Old English Herbals* (Minerva Press, London, 1972)
Thomas, R.E. (1981) 'The Use of Procaine in Geriatrics: A Survey of the Literature', *Aust. J. Pharm. Sci. 10*, 89
Thompson, J.A. *Free Phosphorus in Medicine with Special Reference to its Use in Neuralgia: a Contribution to Materia Medica and Therapeutics* (Lewis, London, 1874)
Walsh, D. *Quacks, False Remedies and the Public Health* (Baillière, Tindall & Cox, London, 1909)

1. Cockayne, O. *Leechdoms, Wortcunning and Starcraft of Early England*, Longman Green, London (1864)
2.

Ƿɪf ſeo ðe menᵹð peſɪeſ ſæð on hɪne mete ⁊ þone þɪcᵹð · ꝥ heo þam pæpneð man þe leoſne ſɪᵹ · ſæſte heo · ɪɪɪ · pɪnᴜeɲ.

In translation this proves to be less than scandalous: 'If a woman mixes pepper seed in her food and thereby hopes to be the more attractive to a man, let her fast for three years.' It is, in fact, a penance.

3. Medvedev, M.A. quoted by Popov and Goldwag (see above)
4. Kim, C. *et al.* (1976) *Amer. J. Clin. Med. 4*, 163
5. Siegel R.K. (1979) *J. Am. Med. Assoc. 241*, 1614
6. Anon (1905) *A Post Office Exposure: Fraudulent use of the U.S. Mails, J. Am. Med. Assoc.* 1391
7. Aslan, A. (1956) *Therapiewoche, 7*, 14
8. Bailey, H. *Will it Keep You Younger Longer?* (Bantam Books, New York, 1977)
9. Berryman, J.A.W. *et al.* (1961) *Brit. Med. J. ii*, 1683
10. Fee, S. and Clark, A. (1961) *Brit. Med. J. ii*, 1680
11. Hirsch, J. (1961) *Brit. Med. J. ii*, 1684
12. Anon (1957) 'Bureau of Investigation Report', *J. Am. Med. Assoc. 165*, 695
13. Danowski, D. *et al.* (1960) *Endocrinology 66*, 788
14. Brearley, R.L. and Forsythe, A.M. (1978) *Brit. Med. J. ii*, 1748
15. Czajka, P. *et al.* (1978) *J. Tennessee Med. Assoc. 71*, 747
16. Presto, A.J. and Muecke, E.C. (1970) *J. Am. Med. Assoc. 214*, 591

Chapter 5

Ritchie, J.M. The aliphatic alcohols. in: *The Pharmacological Basis of Therapeutics*, A.G. Gilman, L.S. Goodman & A. Gilman, eds, 6th Edn, p. 376 (Macmillan, New York, 1980)
Robertson, J. (1806) *A Practical Treatise on the Powers of Cantharides* (Mundell, Doig & Stevenson, Edinburgh)
Russell, J.F. (1973) 'Methaqualones "Heroin for Lovers" ', *Miss. State Med. Assoc. 14*, 496
Thompson, C.J.S. *The Mystic Mandrake* (Rider, London, 1934)

1. Tardieu, A. *L'Empoisonnement* (Paris, 1875)
2. Duhren, E. *Der Marquis de Sade und Seine Zeit* (Zaufl, Berlin, 1906)
3. Ewart, W.B. *et al.* (1978) *Can. Med. Assoc. J. 118*, 1199
4. Chen Ruiting *et al.* (1980) *China Med. J. 93*, 183
5. Lecutier, M.A. (1954) *Brit. Med. J. ii*, 1399
6. Meynier, J. (1893) *Arch. Med. Pharm. Milit. 22*, 53
7. Craven, J.D. and Polak, A. (1954) *Brit. Med. J. ii*, 1386
8. Inaba, D.S. *et al.* (1973) *J. Am. Med. Assoc. 224*, 1505
9. Matthew, H. *et al.* (1968) *Brit. Med. J. ii*, 101
10. Pacht, A.R. and Cowden, J.E. (1974) *Crim. Justice & Behav. 1*, 13
11. Taberner, P.V. (1980) *Psychopharmacology 70*, 283
12. Masters, W.H. and Johnston, V. *Human Sexual Response* (Little, Brown, Boston, Mass., 1966)
13. Beckman, L.J. (1979) *J. Stud. Alc. 40*, 272
14. Rubin, H.B. and Henson, D.E. (1976) *Psychopharmacology 47*, 123
15. Ylikahri, R. *et al.* (1974) *J. Steroid Biochem. 5*, 655
16. Kinsey, B.A. *The Female Alcoholic: A Social Psychological Study* (Charles C. Thomas, Springfield, Ill., 1966)
17. ICCA 30th International Institute on the Prevention and Treatment of Alcoholism, Athens (1984)

268   *Bibliography*

Chapter 6

Clinical Neuroendocrinology, L. Martini & G.M. Besser, eds. (Academic Press, New York, 1977)

Haire, N. *Rejuvenation* (George Allen & Unwin, London, 1924)

Kinsey, A.C. *et al. Sexual Behaviour in the Human Male* (Saunders, Philadelphia, 1948)

Kinsey, A.C. *et al. Sexual Behaviour in the Human Female* (Saunders, Philadelphia, 1953)

Perry, J.S. (Ed) Effects of Pharmacologically Active Substances on Sexual Function: Proceedings of the Second Symposium of the Society for the Study of Fertility, Exeter, July 1967. *J. Reprod. Fert. Suppl. 4* (1968)

Sandler, M. and Gessa, G.L. *Sexual Behaviour, Pharmacology and Biochemistry* (Raven Press, New York, 1975)

Sigell, L.T. (1978) 'Popping and Snorting Volatile Nitrites: A Current Fad for Getting High', *Am. J. Psychiat. 135*, 1216

1. Brown-Sequard, C.E. (1889) *Lancet II*, 105
2. Kolodny, R.C. *et al.* (1971) *New Engl. Med. J. 285*, 1170
3. Cooper, A.J. *et al.* (1970) *Brit. Med. J. iii*, 17
4. Salmon U.J. *et al.* (1943) *J. Clin. Endocrinol. 3*, 235
5. Bommer, J. *et al.* (1979) *Lancet ii*, 496
6. Cooper, A.J. *et al.* (1972) *Brit. J. Psychiat. 120*, 59
7. Malamud, W. and Sands, S.L. (1947) *Am. J. Psychiat. 104*, 231
8. Henson, D.E. and Rubin, H.B. (1978) *Behav. Res. Ther. 16*, 143
9. Griffit, W. (1975) *Arch. Sex. Behav. 5*, 403
10. 'I will please the lord in the land of the living'. Psalms 116:9
11. Haas, H. *et al.* (1963) *Psychopharmac. Serv. Cent. Bull. 2*, 1
12. Forsberg, L. *et al.* (1979) *Fert. Steril. 31*, 589
13. Frank, E. *et al.* (1978) *New Engl. J. Med. 299*, 111
14. Levin, R.J. and Wagner, G. (1980) *J. Physiol. (Lond.) 302*, 22p
15. Persky, H. *et al.* (1978) *Arch. Sex. Behav. 7*, 157
16. Bancroft, J. *et al.* (1983) *Psychosom. Med. 45*, 509
17. Schreiner-Engel, P. *et al.* In: *Proceedings of the 5th World Congress of Sexology* Hoch, Z. and Leif, H.I. (eds) (Excerpta Medica, Amsterdam, 1981)

Chapter 7

Barnes, T.R.E. *et al.* (1979) 'Psychotropic Drugs and Sexual Behaviour', *Brit. J. Hosp. Med. 21*, 594

Buffum, J. (1982) 'Pharmacosexology: The Effects of Drugs on Sexual Function', *J. Psychoactive Drugs 14*, 5

Gay, G.R. *et al.* (1981) 'Love and Haight: The Sensuous Hippie Revisited', *J. Psychoactive Drugs 14*, 111

Grinspoon, L. & Bakalar, J.B. *Cocaine. A Drug and Its Social Evolution* (Basic Books, New York, 1976)

Halikas, J. *et al.* (1982) 'Effects of Regular Marijuana Use on Sexual Performance', *J. Psychoactive Drugs 14*, 59

Hollister, L.E. (1975) 'Drugs and Sexual Behaviour in Man', *Life Sciences 17*, 661

Jones, E. *The Life and Work of Sigmund Freud*, L. Trilling and S. Marcus, eds, Chapter 6 (Penguin Books, Harmondsworth, 1961)

Lewin, L. *Phantastica. Narcotic and Stimulating Drugs: their Use and Abuse* (Kegan Paul, Trench, Trubner, London, 1931)

Stafford, P. *Psychedelics Encyclopedia* (And/Or Press, Berkeley, California, 1977)
Sykes, W.S. *Essays on the First Hundred Years of Anaesthesia* (Livingstone, Edinburgh, 1960)

1. Parr, D. (1976) *Brit. J. Addict. 71*, 261
2. Bell, D.S. and Trethowan, W.H. (1961) *Arch. gen. Psychiat. 4*, 74
3. Shader, N.I. (Ed.) *Psychiatric Complications of Medical Drugs* (Raven Press, New York, 1972)
4. Jones, R.T. (1971) *Pharmacol. Rev. 23*, 359
5. Moreau, J.-J. *Hashish and Mental Illness*, H. Peters and G.G. Nahas, eds (Raven Press, New York, 1973)
6. Kolodny, R.C. *et al.* (1974) *New Engl. J. Med. 290*, 872
7. Kolansky, H. *et al.* (1971) *J. Am. Med. Assoc. 216*, 486
8. Goode, E. (1972) *Am. J. Psychiat. 128*, 1272
9. Wesson, D.R. (1982) *J. Psychoactive Drugs 14*, 75
10. Siegel, R.K. (1982) *J. Psychoactive Drugs 14*, 71
11. d'Orta, G. (1563) *Coloquios dos simples, e drogas he cousas medicinas de India*, quoted by Guerra (see below)
12. Guerra, F. (1974) *Brit. J. Addict. 69*, 269
13. Cushman, P. (1972) *NY State J. Med. 72*, 1261
14. Parr, D. (1976) *Brit. J. Addict. 71*, 261
15. Israelstam, G. *et al.* (1978) *Brit. J. Addict. 73*, 319
16. Mudd, J.W. (1977) *Am. J. Psychiat. 134*, 922
17. Gay, G.R. *et al.* (1982) *J. Psychoactive Drugs 14*, 111
18. Green, R.C. (1959) *Am. J. Med. Assoc. 171*, 1342
19. *UN Bulletin on Narcotics, XXXII* (3) (1980)

Chapter 8

Buffum, J. (1982) 'The Effects of Drugs on Sexual Function. A Review', *J. Psychoactive Drugs 14*, 5
Cooper, A.J. (1972) 'Diagnosis and Management of Endocrine Impotence', *Brit. Med. J. ii*, 34
Mann, T. (1968) 'Effects of Pharmacological Agents on Male Sexual Functions', *J. Reprod. Fert. Suppl. 4*, 101

1. Holmberg, G. and Gershon, S. (1961) *Psychopharmacologia 2*, 93
2. Margolis, R. and Leslie, C.H. (1966) *Curr. Ther. Res. 8*, 280
3. Margolis, R. *et al.* (1967) *Curr. Ther. Res. 9*, 213
4. Miller, W.W. (1968) *Curr. Ther. Res. 10*, 354
5. Sobotka, J.J. (1969) *Curr. Ther. Res. 11*, 87
6. Cooper, A.J. *et al.* (1973) *Irish J. Med. Sci. 142*, 155
7. Mortimer, C.H. *et al.* (1974) *Brit. Med. J. iv*, 617
8. Benkert, O. *et al.* (1975) *Neuropsychobiology 1*, 203
9. Ambrosi, B. *et al.* (1977) *Clin. Endocrinol. 7*, 417
10. Jenkins, R.B. *et al.* (1970) *Lancet ii*, 177
11. Bowers, M.B. *et al.* (1971) *Am. J. Psychiat. 127*, 1691
12. Brown. E. *et al.* (1978) *Am. J. Psychiat. 135*, 1552
13. Goodwin, F.K. *et al.* (1970) *Lancet i*, 908
14. Korten, J.J. *et al.* (1973) *J. Am. Med. Assoc. 226*, 355
15. Laver, M.C. (1974) *Aust. New Zeal. J. Med. 4*, 29
16. Alexander, W.D. and Evans, J.I. (1975) *Brit. Med. J. ii*, 501
17. Smith, B. and Prockop, D.J. (1962) *New Engl. J. Med. 267*, 1338

18. Oswald, I. (1966) *Brit. J. Psychiat. 112*, 391
19. Doust, J. *et al.* (1972) *J. Nerv. Ment. Dis. 155*, 261
20. Sicuteri, F. and Del Bene, E. (1975) *Acta. vit. enzymol. 29*, 100
21. Cremata, V.Y. (1966) *Clin. Pharm. Ther. 7*, 768
22. Feldman, J.M. *et al.* (1977) *Horm. Metab. Res. 9*, 156
23. Benkert, O. *et al.* (1976) *Arzneim. Forsch. 26*, 1369
24. Beaumont, G. (1973) *J. Int. Med. Res. 1*, 469
25. Eaton, H. (1973) *J. Int. Med. Res. 1*, 432
26. Freed, E. (1983) *Med. J. Aust. i*, 551
27. Zacest, R. *et al.* (1983) *Lancet i*, 1221
28. Rubin, H.B. *et al.* (1979) *Behav. Res. Ther. 17*, 305
29. Barbeau, A. (1969) *Can. Med. Assoc. J. 101*, 59

Chapter 9

Comfort, A. (1971) 'Likelihood of Human Pheromones', *Nature (Lond.) 230*, 432

King, J. (1983) 'Have the Scents to Relax', *World Medicine 19*, 29

Rogel, M.J. (1978) 'A Critical Evaluation of the Possibility of Higher Primate Reproductive and Sexual Pheromones', *Psychol. Bull. 85*, 810

1. Curtis, R.F. *et al.* (1971) *Nature (Lond.) 232*, 396
2. Waltman, R. *et al.* (1973) *Lancet ii*, 496
3. Morris, N.M. and Udry, J.R. (1978) *J. Biosoc. Sci. 10*, 147
4. *Perfumery* (Creative Leisure Series, International Publications Service, New York, 1975)
5. Black, S.L. and Biron, C. (1982) *Behav. Neural. Biol. 34*, 326
6. *The Role of Pleasure in Behaviour*, R.G. Heath, ed. (Harper & Row, London, 1964), p. 219

INDEX

272   *Index*